MANAGING CONSTRUCTION CONTRACTS

JOHN WILEY & SONS, INC.
CONSTRUCTION BUSINESS AND MANAGEMENT LIBRARY

COST ENGINEERING FOR EFFECTIVE PROJECT CONTROL
Sol A. Ward

MANAGING CONSTRUCTION CONTRACTS:
Operational Controls for Commercial Risks, Second Edition
Robert D. Gilbreath

MANAGING CONSTRUCTION CONTRACTS
Operational Controls for Commercial Risks

Second Edition

ROBERT D. GILBREATH

JOHN WILEY & SONS, INC.
New York / Chichester / Brisbane / Toronto / Singapore

Copyright © 1992 by John Wiley & Sons, Inc.

Library of Congress Cataloging in Publication Data:

Gilbreath, Robert D. (Robert Dean), 1949–
    Managing construction contracts : operational controls for
commercial risks 2nd edition / Robert D. Gilbreath.
        p. cm.
    ''A Wiley-Interscience publication.''
    Includes index.
    ISBN 0-471-56932-1
    1. Construction industry—Management. 2. Construction contracts.
3. Risk management. I. Title.
HD9715.A2G49 1992
624'.068—dc20   91-32241
                                                                      CIP

Printed and bound in the United States of America by Braun-Brumfield, Inc.

10 9 8 7 6 5 4 3 2 1

For Linda, Bob, and Alice

Special thanks are in order for my attorney, Henry W. Quillian, III, for his advice and assistance, and for graciously contributing Chapter 20, *The Role of a Lawyer in Contract Management.*

PREFACE

There are two ways to look at a construction project. You can visualize it in terms of steel, concrete, cables, and pipe—the technical view. Or you can see what most people do not—the complex set of roles and relationships without which nothing would be built. This is a book about those roles and relationships, those contracts—how they are planned, formed, administered, and managed; how the participants are selected and how they deal with each other so that the project is a commercial success for everyone concerned; and how they avoid or resolve the inherent and special difficulties that each faces.

Special difficulties abound in construction. By its very nature, this business is fraught with risk. People and companies are brought together, often for the first time, working in unique combinations, under different agendas, and sometimes with incomplete information and opposing objectives—and all under pressures of time, money, and technology. Risk is the underlying presence, and how you confront risk—recognize it, provide for it, and counter it when you can—is the major factor in whether or not you will be successful.

So the book is focused on risk, not from any particular point of view, but from every perspective—because contracting is not a zero-sum game, nor is it analogous to war. With those, you win when the other party loses. With contracting, whether you are the owner, contractor, design professional, or anyone else involved, you usually lose when the other party loses, one way or the other.

I take this approach because I believe that good contract management is in everyone's interest. The object is to get high-quality work on time and at a fair price, not to make or defeat adversaries. Of course, sometimes it comes to that, and I've got some advice in those instances. But the overriding objective is to make the commercial risk manageable and the contract experience satisfying.

Because the owner or buyer is ultimately responsible for all phases of the work and benefits or suffers accordingly, I address contract management from that perspective. But the book and the methods and controls it presents apply to everyone with an interest in commercial success. Where these coincide, so much the better. Where they diverge, I point out the differences.

The scope of our concern is everything that makes or breaks the contract experience. We start with critical decisions made (or avoided!) early, during the strategic phase of planning. These include the overall project organization, the roles and relationships among the players, and the pricing and timing of contracts. Then we will proceed with the formation process: how bidders are selected, bids solicited and received, and contracts written. Administration comes next—the day-to-day processing of payments, changes, claims, and acceptances. Finally, I present some recommendations for sound contract reporting and auditing to assure that the entire process of contract management is operating as it should.

Along the way you'll find guidelines for making choices and tools to implement them once you have. Choices and chances make up the business of contract management. There are no "right" ways to do anything in this field, only ways that make sense for you, under your circumstances, to achieve your goals. So I present choices to you: ways to organize, price, resolve conflicts, manage risk, and so on. You learn what is involved with each alternative so that you can make the proper selection and, most important, be prepared for the responsibilities and risks it entails.

Since the first edition was published I have taught contract management at more than 100 training sessions on six continents. I have listened to a lot of horror stories along the way and picked up dozens of new tricks and refreshing ideas from thousands of readers and attenders. You will find many of them here. You will also find changes that reflect the evolving nature of the construction industry and the discipline of contract management.

Accelerating political, economic, technical, and cultural change has not missed any business, especially this one. Chief among these for contract management might be technology: the ubiquity of personal computers, workstations, and telecommunications systems. Paperwork and record keeping were once the bane of contracting—but no longer. Now we can do it faster, with less error—and we can do it anywhere we want! Information technology has made moot the old arguments between home office and field, between headquarters and job site. It has also torn down the organizational wall between those who plan and form contracts and those who administer them. And it has given the power of control to the smallest of projects.

International work is much more common, as the economies of the world and the endeavors of corporations knit themselves together in webs of interdependence. So I have added a special section on the perils and thrills of cross-border contracts. More exotic, more rewarding, perhaps, but undoubtedly more risky. Knowing those risks is vital, because sooner or later all of us are going to be working across borders and between cultures.

Perhaps the most welcome change, to me at least, has been the gradual acceptance of contract management and contract control techniques around the world. Contract specialists, contract administrators—whatever they are called—don't seem so rare anymore. And the need to plan, form, administer, and manage contract work is not questioned as a discipline demanding special insights, skills, and tools. So while I have spent a number of years and a ton of patience explaining why this subject is important, I do not have to convince many people today. Now I can focus not on *why* but on *how*—how best to take up the challenge of contracts. The people and companies that needed to be told *why* this is important aren't around anymore. They either bought into the idea or they went bankrupt. Failure is a stern teacher.

Actually, this whole question of *who?* seems to be irrelevant today, thankfully. Who should write contracts? Who should sign change orders? Who decides whether a payment should be made? Modern organizations are eliminating the barriers between departments and disciplines, as they should. The answer to the *who?* question is: you! You should be concerned with how the contracts are written, how the estimating and payment process works, what causes unnecessary change orders, and how to prosecute or defend claims. You face the risk, regardless of your position or speciality, and you have a stake in the conflict. Not a conflict among people or companies, but one pitting every person and every entity involved against a more tenacious and powerful enemy—*failure,* the ultimate risk. Learn risk and controls, learn how to use one to counter the other, equip yourself with the knowledge and the tools to win, and you *will* win. So will everyone else, including me.

Duluth, Georgia   Robert D. Gilbreath
January 1992

CONTENTS

List of Figures, / xiii
List of Exhibits, / xv
Introduction, / 1

PART I   CONTRACT PLANNING   15

  1. CONSTRUCTION CONTRACTS: ROLES AND
     RELATIONSHIPS   17
  2. ORGANIZATIONAL AND CONTRACTING STRATEGIES   23
  3. CONTRACT PRICING ALTERNATIVES   37
  4. CONTRACT PACKAGING AND SCHEDULING   59

PART II   CONTRACT FORMATION   67

  5. DEVELOPING CONTRACT DOCUMENTS   69
  6. BIDDER QUALIFICATION AND SELECTION   85
  7. ISSUING REQUESTS FOR PROPOSALS   91
  8. MANAGING THE BID CYCLE   119
  9. BID RECEIPT AND EVALUATION   129
 10. CONTRACT AWARD   139

PART III   CONTRACT ADMINISTRATION   145

11. **MOBILIZATION AND COMMENCEMENT**   147
12. **PROGRESS BILLINGS AND PAYMENTS**   157
13. **CHANGE ORDERS**   179
14. **BACKCHARGES**   197
15. **CLAIMS**   203
16. **SHORT-FORM CONTRACTING**   217
17. **CONTRACT CLOSEOUT**   223

PART IV   CONTRACT MONITORING   233

18. **CONTRACT REPORTING**   235
19. **CONTRACT AUDITING**   247

PART V   SPECIAL RECOMMENDATIONS   259

20. **THE ROLE OF A LAWYER IN CONTRACT MANAGEMENT** 261
21. **CONTRACTING ACROSS BORDERS**   275

GLOSSARY OF TERMS   283

INDEX   293

LIST OF FIGURES

1	Major Phases of Contracting	6
2	Contract Responsibilities	20
3	Organizational Approaches	27
4	Impacts of Contracting Approaches	34
5	Typical Contract Pricing Alternatives	38
6	Features of Pricing Alternatives	40
7	Fixed-Price-Incentive-Fee Graph 1	47
8	Fixed-Price-Incentive-Fee Graph 2	48
9	Fixed-Price-Incentive-Fee Graph 3	49
10	Cost-Plus-Incentive-Fee Graph	52
11	Contract Formation Milestones and Activities	60
12	Sample Contract Formation CPM Network	64
13	Typical Bidding and Contract Documents	73
14	Bid Evaluation Process	130
15	Typical Contract Records	153
16	Contract Progress Payment Process	168
17	Typical Paperwork Flow: Progress Payments and Billings	169
18	Change Order Process for a Large Project Organization	189
19	Changed Work Documents	189
20	Backcharging Process	201
21	Evolution of a Claim	213
22	Contract Reporting Hierarchy	239

LIST OF EXHIBITS

1	Contract Packaging Scope Document	62
2	Sample Invitation to Bid	98
3	Sample Proposal	100
4	Sample Agreement	107
5	Sample General Conditions for Construction Contracts (Index Only)	112
6	Sample Special Conditions	114
7	Sample Addendum	122
8	Bid Evaluation Considerations	132
9	Contractor Submittal Checklist	149
10	Insurance Certificate Log	150
11	Summary of Contract Progress Payment Estimate	170
12	Detailed Contract Progress Payment Estimate	171
13	Notification	184
14	Request for Fair Price Estimate	186
15	Change Order Approval Form	187
16	Change Order	188
17	Notification Log	191
18	Change Order Log	191
19	Changed Work History	192
20	Backcharge	200
21	Contract Closeout Checklist	227

22	Contractor Evaluation	228
23	Contract Evaluation Form	229
24	Monthly Contract Cost Report	240
25	Contract Formation Schedule Report	241
26	Claims Summary	242
27	Change Order Summary	243
28	Backcharge Summary	244

INTRODUCTION

Major construction projects represent some of the largest and most complex undertakings known. When completed, they give testimony to a variety of technological methods, engineering skills and innovations, materials, and a tremendous consumption of resources—time, money, and people's talents. What is often overlooked when viewing a massive structure or an intricate operating facility is the intricate web of commercial transactions, business arrangements, and management challenges that were also required.

Much has been written, proposed, and implemented in the way of management controls over the construction of major projects. These efforts involve attempts to control cost, schedule, and technical performance. Yet projects continue to fail. They fail in terms of cost when budgets must be constantly revised upward and when final cost greatly exceeds original estimates. They fail in terms of schedule when completion dates are extended and when the facility is available for use long after the intended date. And they fail when the facility does not meet its technical requirements or is otherwise unfit for its intended purpose.

The business of construction is performed through contracts. And from the perspectives of both owner and contractor it is a business rapidly becoming fraught with risk and uncertainty. The impact of project failures has caused owners and contractors to reexamine the ways they contract for engineering and construction services. They are searching for ways to improve the processes by which they plan, form, and administer construction contracts and to control commercial risks that threaten project success. This book is intended to help them in that search.

COMMERCIAL PRESSURES IN THE CONTRACTING ENVIRONMENT

Today's construction environment is fast becoming more perilous than ever before. Ever-present cost and schedule overruns, pervasive construction disputes and claims, and an abundance of change orders are symptoms of an inability to control or mitigate commercial risks. The frequency and intensity of these risks are increasing because of a number of factors:

Growing commercial pressures caused by economic factors, environmental concerns, financing costs, and more stringent public accountability

Closer outside scrutiny of management controls and practices

Proliferation of design, construction, and management alternatives

Increasing involvement of governmental and regulatory entities

More frequent use of joint ventures, pooled ownership, and other project-specific corporate combinations

Growth of innovative and complex risk-sharing techniques

Increasing complexity of today's contracts

Competition in the industry

Scarcity of Capital Funds

Globalization

Increased activism of affected interest groups

Rights and Responsibilities

If they are written and administered correctly, construction contracts are no more than an array of rights and responsibilities for all parties concerned. Builders are hired as independent contractors, with emphasis on the word *independent*. The owner has no right, absent specific agreements, to direct, supervise, or manage the work for which the contractor has been engaged. Instead, the owner manages the contractual rights and relationships—the contract duties and obligations. The owner specifies the results and allows the contractor to employ the means and methods it sees fit to achieve them.

The owner's major duties (in addition to paying for the work and getting out of the contractor's way) are:

1. *Making Its Expectations Clear and in Advance.* Letting the bidders and eventual contractor know what it wants, when, and where. Clear technical specifications and drawings are a must, but so are the commercial terms and conditions that accompany them.

2. *Inspecting the Work as it Progresses.* Construction is one of the few operations where the owner is involved in approving or rejecting what it buys at intermediate steps in its production. Inspections should be

timely, noninterfering, and reasonable—and according to contractual requirements.

3. *Accepting the Work When It Meets the Expectations Established in the Contract.* Taking ownership and responsibility for finished work, making payments and releasing the contractor of further responsibility.

The Owner's Contractual Arsenal

In a perfect world, the contractor produces results that meet owner expectations, inspections, and eventual acceptance. In reality, though, the path from expectations to acceptance is riddled with pitfalls, and to counter commercial and technical risks, owners and their representatives have a broad arsenal of contractual protections and management tools. Here is a partial list of commercial controls that should be considered for every contract:

1. *Partial and Final Payments.* Sometimes these are called "carrot and stick" controls. The carrot is money. The stick is not getting the carrot.

2. *Retention.* This is the customary withholding of earned money until the end of the work. Again, construction is fairly unique in this regard. Wouldn't it be nice if we could buy an automobile and keep back part of the purchase price until we're satisfied with it?

3. *Future Awards.* Holding out the potential of future contracts is a valuable control tool.

4. *Inspection.* Physical evaluation of the work according to some prearranged or industry-accepted criteria.

5. *Testing.* Subjecting parts of the work to operating conditions or destructive or nondestructive examination.

6. *Independent Certification.* Having a third party check out the contractor's work and report on its adequacy. Usually, the certification is made by an industry-approved agency.

7. *Guarantee and Warranty.* Anyone who can differentiate between these terms should be given the national semantics award.

8. *Bonds.* Bid, payment, and performance bonds issued by sureties after examining the contractor and/or scope of work.

9. *Reporting.* Periodic and incidence-based reports keep the owner apprised as to what has happened or is about to happen. Hopefully, they lead to failure prevention rather than merely documenting how failure occurred.

10. *Expediting.* Stationing purchasing or contract specialists in the fabrication–shipment process to monitor or assist the contractor in getting the right material and equipment to the site in a timely manner.

11. *Referrals.* Recommendations or warnings to other owners regarding the contractor's performance.
12. *Extra Work.* Adding work (or withholding it) to the contractor's scope, depending on how well it is performing.
13. *Submittals.* Reviewing and approving (or rejecting) contractor shop drawings, fabrication schedules, material samples, management procedures, and the like.
14. *Manufacturing Surveillance.* Putting owner personnel in vendor or contractor shops to witness critical manufacturing operations.
15. *Insurance.* Requiring it of contractors, and buying it for the owner's protection.
16. *Arbitration.* Agreeing to present disputes to third parties for resolution.
17. *Litigation.* The law of the land.
18. *Quality Control, Quality Assurance.* Owner programs, contractor programs, owner specification, and oversight of both.
19. *Incentives.* Bonuses and penalties tied to cost, schedule, or technical performance.
20. *Auditing.* Review of construction operations, commercial transactions, billings, and payments to assure they are adequate and in compliance with the contract.

None of these controls is automatic. They must be agreed upon by both parties, specified in the contract documents, and followed throughout the performance period. Each requires further definition, and none work if they aren't understood or enforced. The business of contract management is to see that this is done. *Contract management is the collection of plans, approaches, systems, and activities required to successfully execute and control the commercial experience of contracting.*

Contract management is, therefore, the way to recognize special risks on your project and counter them with intelligent controls—controls that work, give more benefit than they cost, and are acceptable to both parties. These are "operational controls"—not theories or rules of law or engineering. They are sensible, real-world ways to manage contracts. To choose and implement operational controls for your project, you need to understand each major element of contracting—each "phase" of contract management. Each phase presents special risk and demands special controls.

CONTRACTING ELEMENTS

For owners, the challenge of contractual controls may be put this way: How do I achieve a fair and equitable contract that provides acceptable levels of control, and how can it best be administered throughout the performance period? The answer to this question lies with the recognition that contracting controls are vital to project success and that the stakes are too high to ignore time-proven methods ensuring the owner's protection during all phases of the contracting process.

The contracting process begins with the inception of a need to contract for outside services and a determination of the scope of work and continues through the completion of contracted performance, final payment and acceptance, and formal contract closeout or termination. Whether there will be a single design–build contract or as many as 100 prime contracts, each step contains potential for control or failure.

For each contract, four major elements or phases demand management attention and control (see Figure 1).

Contract Planning. Contracting practices should be planned and organized to reflect:

1. Overall company objectives
2. Project-specific circumstances
3. Technical, schedule, and commercial risks

Contract Formation. Construction contracts should be awarded only when the owner is assured that:

1. They are fair and equitable
2. They contain provisions that protect the interests of all parties
3. They embody performance controls
4. They can be properly administered.

Contract Administration. Adequate procedures, enforceable administration controls, and trained personnel should allow the owner to control, rather than react to, commercial risks and pressures during the performance of contracted work.

Contract Monitoring. Management attention should be focused on the previous three phases through the use of:

1. Timely and informative contract reporting
2. Cost-effective contract auditing

Each of these phases, together with the risks, objectives, and activities they entail, are the subject of chapters in this book. Part 5, Special Recommendations, also contains two chapters of particular interest. In Chapter 20 we describe how to choose and use competent legal counsel, some of the special tasks a lawyer can perform, and when and why that expertise is crucial. Finally, Chapter 21 is designed to point out the special risks and

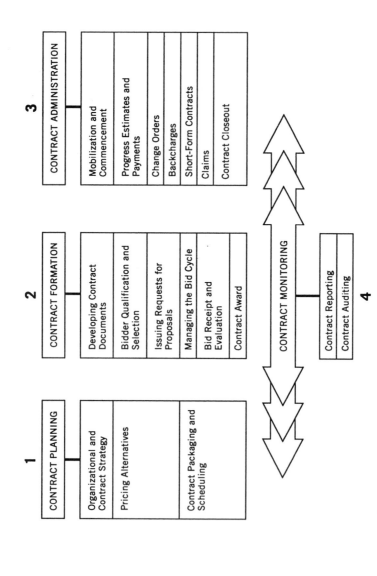

FIGURE 1. Major phases of contracting.

1	2	3
CONTRACT PLANNING	CONTRACT FORMATION	CONTRACT ADMINISTRATION
Organizational and Contract Strategy	Developing Contract Documents	Mobilization and Commencement
Pricing Alternatives	Bidder Qualification and Selection	Progress Estimates and Payments
	Issuing Requests for Proposals	Change Orders
	Managing the Bid Cycle	Backcharges
Contract Packaging and Scheduling	Bid Receipt and Evaluation	Short-Form Contracts
	Contract Award	Claims
		Contract Closeout

CONTRACT MONITORING

4

Contract Reporting
Contract Auditing

controls that apply to contracts which, in one way or another, involve more than one country. A brief description of each contract management phase follows.

CONTRACT PLANNING

Because construction of a major facility is usually secondary to other ongoing owner operations, such as the production and sale of goods or services, owners are generally at a decided disadvantage when entering the world of construction contracting. Before doing so, there is no substitute for thorough planning. This planning should follow a structured, disciplined approach considering the owner's objectives, the existing contracting environment, and alternatives for project execution and control. As such, an owner's contract plan should address the details described in the following subsections.

Project Organizational and Contractual Strategies. Owners should choose an organizational alternative that best suits their control objectives and requires a level of involvement they are able to meet. Options range from a design–build approach, where the owner awards one huge contract for virtually all project-related services, to the multiple-prime concept, where as many as 200 separate contracts might be awarded. In between these two extremes lie a wide variety of approaches, each requiring different control strategies and different levels of owner involvement throughout the project's life. Owners should review each possible approach, analyze its features and benefits, and select the one best suiting their objectives and capabilities.

Contract Pricing Alternatives. There are an infinite number of ways to price and pay for contractual performance, ranging from the straightforward lump-sum method to highly sophisticated cost-sharing techniques. To each may be applied a myriad of cost, schedule, and technical incentives, escalation provisions, and risk-sharing elements. Each has been effective given certain project-specific circumstances, and each has failed given others. The selection of a specific pricing method should be based on a thorough understanding of its features, control benefits, and the environment to which it will be applied.

Contract Packaging and Scheduling. The subdivision of work into contract or subcontract packages and the sequencing of their formation and award activities over time represent a major challenge. Often, contract documents are rushed through the formation cycle without adequate definition of scope, complete design information, and due consideration for project schedule needs. The result is uncompetitive and overpriced bids, overlaps and gaps in scope of work among several contracts, and an uncoordinated contracting

n. Thorough planning and scheduling of each contract process is necessary, along with adequate scope definition and a tailored pricing scheme. Well-conceived contract plans should be implemented before bidding commences, and critical schedule constraints in the design and construction phases should be incorporated into the contract bidding and award plan.

This formalized contract planning is valuable in many ways. It surfaces scope inconsistencies among the anticipated contracts, affords conscientious pricing decisions, and outlines the support information, from the engineering and design efforts, required for contract bidding and award. It helps those responsible for contract formation to organize and plan their functions and construction management personnel load field monitoring resources and support services over the construction period.

CONTRACT FORMATION

Contract formation begins with the decision to contract for project-specific goods and/or services and ends with a written agreement signed by both parties—the contract. Poor bidding and contract documents, inadequate bidder qualification and bid management, and poor negotiation efforts are damaging in themselves, but their greatest damage is to contract performance and administration efforts that follow. They result in selection of the wrong contractor, often at the wrong price, and under the wrong terms. Control over the formation process involves the steps discussed in the following subsections.

Developing Contract Documents. If you want it, get it in the contract: this advice has never been more warranted than it is today. A good set of contract documents is clear and enforceable, the intent of the contracting parties is without question, and the interests of all parties are well protected. Most owners spend considerable effort in designing and adapting "reference" or "standard" contract documents. They should not only represent the highest standards of legal practice but also reflect sound engineering and construction judgment and include effective management controls. Standard or reference documents, each tailored to a particular application, lessen the possibility of misunderstanding, undue compensation, and change orders, claims and litigation.

Bidder Qualification and Selection. Unless restricted by regulations or funding requirements, owners are free to choose prospective contractors from whom to receive bids. These should be prequalified to rigid quality, financial, and performance standards before being allowed to submit bids. Proper bidder qualification helps reduce the bid evaluation effort by eliminating those prospective bidders who prove unacceptable. It helps ensure contractor stability, avoiding performance problems should an unqualified contractor receive an award. And it promotes the widest solicitation of qualified bids, thereby increasing competition among bidders and driving prices downward.

Issuing Requests for Proposals. When preparing contract bid requests, owners can protect their interests in two ways. First, they can disseminate all available information to bidders, allowing for competitive, precise bidding based on facts rather than assumptions or contingency considerations. Second, they can channel the incoming bids into a structure that aids speedy bid evaluation, reduces postbid clarifications and negotiations, and ensures that fair and workable terms and conditions are incorporated into the final contract documents.

Managing the Bid Cycle. Selecting the best contractor under the best terms and conditions and at the right price begins long before bids are received. During the bid cycle—the time from release of requests for proposals to prospective contractors until bids are received—owners can take steps that affect the quality and price of bids. These include clear and timely modifications to the requests for proposals (addenda), adequate site visits, should they be required, and controlled and coordinated prebid meetings. Each of these steps is designed to give bidders clear and consistent information and allow them to estimate prices precisely—without unwarranted padding for perceived risk and contingency.

Bid Receipt and Evaluation. Once bids are received, two interrelated objectives should be met: (1) a complete and impartial review should lead to selection of the best offer, and (2) the confidentiality of bid and bid evaluation information should be maintained. Evaluation should consider not only first costs, that is, the cost to have the work performed, but also the costs of operating and maintaining the finished work in the future. These "life-cycle" costs often result in the owner selecting a bidder who, though not the lowest, will provide the least expensive product in the long term.

Bid security measures should be initiated before bids are received and followed throughout the evaluation process. Failure to secure confidential information may result in reluctance of contractors to bid on future work (thereby reducing competition), compromised negotiation postures, and starting a contractual arrangement on a less than good faith basis.

Contract Award. As soon as a contractor has been chosen, the contract documents that will govern the arrangement should be prepared and executed by both parties. This generally involves incorporation of bidder-specific information (prices, schedules, and so on) into the contract documents, as well as the accommodation of all addenda issued during the bid cycle. Quite often other postbid agreements are also documented in the contract, such as the results of clarifications and negotiations. It is not uncommon for owners to delay this process and instruct the contractor to commence work under "letters of intent" or similar agreements, pending conformance of the actual contract documents. Letters of intent typically lack performance controls and leave all parties exposed to commercial risks.

CONTRACT ADMINISTRATION

Contract administration is a term describing the commercial handling of a contract once it is awarded and until it is formally terminated (contract closeout) or dies an unplanned death (through contract default or early termination). Contract administration controls ensure commercial compliance with the terms of each contract. Even under ideal conditions, this is generally an arduous task for project owners. They must monitor not only the compliance by contractors but also their own compliance. Routine duties involving receipt and review of contractor submittals, maintaining contract records, and monthly progress payments sometimes obscure critical risks encountered during the performance period. Change orders, claims, early payment or overpayment, and unsatisfactory performance can be prevented by implementing a structured contract administration program.

Mobilization and Commencement. The initial phases of contract administration set the stage for success or failure. Steps taken to begin a good faith relationship, one that will prevail throughout the performance period, include formal commencement meetings between all parties, covering the commercial transactions they may encounter. In addition, this period requires establishing complete, protected, and easily accessed contract files. The mountain of contractual records, documents, forms, and reports needed demands careful planning and discipline. Not only are these essential for contract administration, but considering the fact that they could be used for claims or lawsuits years later, this documentation becomes even more critical.

Progress Billings and Payments. Unlike the purchase of finished goods, the buying of contract services requires periodic progress payments as the work evolves. Unplanned or subjective determination of progress results in overpayments or underpayments, loss of control, and continuous disputes. This can damage the owner in two ways: in view of the cost of financing, owners who pay for performance early incur unnecessary costs due to the time value of money; and once payment has been made for performance yet to be accomplished, owners lose payment leverage.

Change Orders. Virtually every contract changes to some extent before it is completed. Formal and constructive changes are the most pervasive and threatening factors jeopardizing successful project completion. A change-control program that identifies changes early, allows for thorough evaluation of the need for a change before it is ordered, and provides responsible pricing and payment controls is essential. Changes should be controlled on a selective basis that allows management flexibility to accommodate unforeseen circumstances, yet provides a minimum degree of control at higher levels. Balancing the risk of improper change authorization with the freedom to implement management decisions in a timely manner is a major challenge.

Backcharges. Charging a contractor for the cost of correcting inadequate work or work that was not performed represents a reversal of normal owner–contractor roles. Many owners fail to collect these costs by not notifying the offending contractor or vendor before undertaking the corrective work, or through inadequate control of costs or cost records. As a result, they often pay two or three times what they should. Aggressive backcharging, where appropriate, can represent substantial savings to owners in this position. The potential for backcharges increases as the number of prime contractors and vendors involved increases.

Short-Form Contracting. All contracting needs and transactions cannot be anticipated months or years in advance and incorporated into contractual provisions. Certain efforts almost always take place on a local, or project-site, level. Generally, these involve short-form contracts for short-duration, low-cost work and field-office-issued change orders to accommodate emergency or unforeseen events. Judicious use of these methods represents a sound control policy that allows responsive reaction to events that inevitably occur during the construction period. However, the necessary control over contract formation and administration should not be abandoned for the sake of brevity or expediency.

Claims. Claims are not standardized, but they do follow general characteristics, stem from similar circumstances, and have predictable elements. They are also quite common, so it pays to prepare for the defense or pursuit of claims. The best claims defense will always be to avoid claim-provoking conditions through proper contract formation and administration. When claims are unavoidable, every effort should be made to identify claims situations early, thoroughly analyze each claim, and press for rapid resolution. Prolonged disputes jeopardize remaining project performance and create unnecessary and bothersome litigation, generally to the detriment of all parties involved.

Contract Closeout. Whether terminated in a planned or unplanned fashion, each contract should be formally "closed out." The process should be structured to protect the owner's interest by ensuring (1) that all performance is acceptable (or otherwise noted), (2) that all expected documentation is received and adequate, (3) that any other deliverables have been received, and (4) that no damaging postperformance actions will take place. This usually involves the joint effort of many people on the part of the owner, and their efforts should be coordinated so that final payment is not made while any element of performance is lacking. In addition, information regarding each contractor's performance should be used for qualifying potential contractors for future work.

CONTRACT MONITORING

There are two major tools which help to ensure that controls needed for proper contract planning, formation, and administration are in place and operating. Every party should use both. They are: (1) contract reporting, and (2) contract auditing.

Contract Reporting. Management needs to know about contracting status on both a continuous and exception basis. Periodic reports should identify key performance factors and identify and describe such threatening elements as claims, change orders, backcharges, default, as well as cash requirements to support the construction effort. A system should provide for exception reporting and allow traceability of problems to their source—allowing management to evaluate and take corrective action in a timely manner.

Contract Auditing. Two distinct types of auditing are important to assist owners in ensuring that their interests are being protected. Cost audits allow you to verify that charges subsequent payments are accurate, in accordance with the contract provisions, and represent work that was actually performed. The audit approach and work program should be tailored to each specific contract type and focus on major elements or risks regarding error or overcharge. Operational audits, on the other hand, are forward looking in perspective. They give management insight into the efficiency and economy of contracting practices. When properly structured and targeted, operational audits can help eliminate unnecessary risk, streamline operations, and ensure consistency of contractual controls.

BENEFITS OF CONTRACT MANAGEMENT CONTROLS

The controls described in this book are intended to make the contract experience a satisfying one for every party involved by lessening the risk of commercial failure during contract planning, formation, and administration. Contract planning considers commercial risk and allows owners to select the number, scope of work, and pricing structures for contracts that best meet company objectives.

Formation controls lead to contracts with capable contractors at the lowest possible price. Well-designed documents ensure commercial compliance and reduce the formation effort. Negotiating effort and evaluation costs are also reduced. Scope of work is well defined, decreasing the likelihood of damaging claims, litigation, and costly change orders.

During the contract performance period itself, prepayment and overpayment are avoided. Payment leverage is maintained and helps to ensure adequate performance. Change orders are dramatically reduced in number and in impact, and only those changes that are needed are made—with full knowledge of the cost and effect of each in advance. The frequency and cost of claims is lessened, and they are resolved swiftly and fairly. Backcharge opportunities are surfaced and pursued whenever possible.

Informative contract reporting lets us focus corrective action when necessary. Contract audits bring direct cost savings, in many cases, and long-term indirect savings when operating and system deficiencies are corrected.

Each of these controls is designed to achieve the ultimate objective: a quality project on time and within budget. When taken together, they represent a sound program of contract management.

SCOPE OF THIS BOOK

Some assumptions must be made to place the suggestions and assumptions used here in the proper context.

First, the contracting processes will be assumed to be private; that is, there will be little if any discussion of government or public contracting processes and their commensurate regulations, procedures, or contracting agencies. Other than the guidelines under which private contracting occurs, such as contract law, industry practices, and commercial expedience, the descriptions and recommendations presented are governed by sound principles of engineering judgment, accepted project management techniques, and business acumen. Reliance on published regulations for particular contract environments is avoided. In their absence, a pragmatic approach is taken to determine "what works best for all concerned."

Associated topics such as contract terminology, or phraseology, case law, arbitration processes, and the like (acceleration, warranties, hold harmless clauses, legal liability, suretyship, and so on) are not the subject here. Instead, we deal with operational aspects of construction contracting—how a contract is best achieved and administered.

Of course, contract activities cannot be discussed in a vacuum. They are extremely dependent upon a multitude of interrelated functions and will vary with particular project, organizational, and industry settings. For this reason, we will restrict ourselves to commercial or industrial projects in the public and private sectors, ones that require extensive engineering and design activity.

Forms, documents, and reports that are included are intended as examples. Evaluate them carefully before adapting them to your contract management uses.

Although the guidelines and suggestions presented here are often written from an owner's perspective, they work in favor of all project participants, including contractors, engineering firms, consultants, construction managers, and so on, in achieving a common goal—successful project completion.

PART I

CONTRACT PLANNING

Before the first steps of the contract formation and administration processes are taken for a major project, several critical decisions must be made concerning the content and management of construction contracts to be employed. This contract planning is an integral part of a broader overall project management plan. Where the project plan would address the project schedule, anticipated cost, technical requirements, and responsibility assignment for the various project participants, the contract plan deals with the contracts to be employed to perform the work in accordance with that project plan. The project plan seeks to answer the questions, What will be done, and when will it be done? Contract planning deals with, Who will we hire to do it, and how will we ensure that it is done at the most economical costs, with the least risk to us? Contract planning, then, seeks to outline the methods by which the business of construction will be conducted and controlled.

Elements of a good contract plan include:

Organizational and Contracting Strategies. These decisions concern the delegation of work to project participants and are major considerations of the owner's project management philosophy. Depending on the alternatives chosen for a project, the resulting impact on contract functions will vary widely. We will examine the project organizational approaches used for construction from this point of view—how each choice affects owner risk and needed controls over contract formation and administration.

Contract Pricing Alternatives. There are an infinite number of ways to price and pay for construction work, and each has succeeded in ensuring performance in some environments though failing in others. We will examine the nature of available pricing alternatives in order to understand the features of control (or risk) inherent to each and to discern the need for various formation and administration controls, selectively applied, depending on alternative pricing structures for construction contracts.

Contract Packaging and Scheduling. The number, respective scopes of work, design requirements, and sequencing of formation activities for each contract are choices that must be made during a project's planning phase. We will present key considerations for the selection of each while addressing the interfaces of contract formation and administration functions with overall project scheduling and coordination objectives.

In this section we explore these three major topics and relate them to the practices of contract formation, administration, and monitoring, described in later chapters.

CHAPTER 1

CONSTRUCTION CONTRACTS: ROLES AND RELATIONSHIPS

Most projects require that many contractors, subcontractors, material suppliers, manufacturers, and others carry out the functions that traditionally have not been performed by the engineer, architect, or project owner. Whether the project is an electric generating station, pipeline system, chemical plant, office building, or manufacturing facility, these organizations perform various activities, undertaken for consideration (money) from the owner, and are assigned various duties and responsibilities as well as rights. These elements are commonly reduced to written form and represent a contract between the owner (buyer) and the contractor (seller).

CONTRACT DOCUMENTS

The contract documents are the script, then, by which the parties involved perform. As such, their importance cannot be overstated. Contract documents (the "contract") will be discussed in greater detail in Chapter 5. For now, however, we must understand that a construction contract is merely a set of criteria, or expectations, that bind the contracting parties. Construction contracts do not normally contain detailed instructions or step-by-step procedures covering all the activities to be performed. In almost all cases, they are not perfect, exact, or ideal. They are written and interpreted (or ignored) by human beings and applied to complex, high-stakes environments subject to the inadequacies of organizations and the whims of events that are, at times, out of their control, such as weather, strikes, natural disasters, and labor and resource constraints.

The inherent difficulties of construction cause us to control what is controllable and provide for risks associated with those events beyond our control. This is the purpose of contract management.

FORMATION VERSUS ADMINISTRATION

Contract formation and *administration* are used here to represent the business management of construction contracts—that is, contracts under which construction site labor is expended as a part of the scope of performance. Included are contracts where the performer (contractor) provides material, equipment, and labor (furnishing and installing process piping, for example) or only labor and its tools (such as under a contract for foundation excavation).

These twin functions, formation and administration, are generally divided into two chronological categories: those performed up to and including the time a construction contract is established (contract formation), and those that continue throughout the performance period, ending, it is hoped, with successful completion of the contracted work, final payment, and closeout—called contract administration. Put simply, contract formation is the process of securing a written and signed contract, whereas administration is the handling of the contract until it expires, through natural causes or otherwise.

THE CONTRACT MANAGER

To avoid semantics problems and organizational slants, we will assume that one person, or organization, represents the owner in seeing that contract formation and administration activities are performed. For lack of a more accurate title, we will use the generally accepted term *contract manager* in referring to such a person or organization. As we will see, contract formation and administration are eclectic functions, interfacing with many disciplines and traversing many internal and external organizational boundaries. As such, the contract manager must possess an eclectic collection of project-related skills and be acutely aware of the needs, methods, and products of other project participants. Contractors, of course, will use contract managers to perform contract management from their perspective. Whether either party designates a specialist to do this work is not important—just that it be done.

As risk- and capital-intensive projects become commonplace around the globe, spanning many political, organizational, and management orientations, a host of relatively new construction-related specializations, such as contract management, have emerged to cope with these technical and commercial complexities. Twenty years ago professionals claiming to be cost engineers, risk managers, quality assurance specialists, construction specifiers, or project control specialists would have found it difficult competing for recognition among organizations engaged in construction projects. For those who perform contract management duties, this has been a difficult task for two reasons:

1. Contract management is a relatively new function and as such must impinge on real or imagined professional territory collectively occupied by other, more established and recognized participants on the construction project management team. Jurisdictional disputes involving authority, responsibility, and interface relationships are common. This condition is particularly aggravated by the high degree of visibility contract management enjoys among financial and legal management.

2. There is no generally accepted definition of contract management or of the qualities and skills required. To practice good contract management, we must first achieve a common understanding as to what, in fact, it is.

Performance Monitoring Versus Contract Administration

To operate at arm's length with a construction contractor and maintain the independent contractor relationship, the project owner and its representatives cannot exercise direct control over the means and methods employed to construct the work. These are left for the contractor to control. The contractor will provide the labor, material, and equipment necessary to fulfill the commitment, and is also solely charged with supervising that the work is performed adequately.

But the owner, on the other hand, has the right to expect that the project is being designed and built according to its requirements and established standards of quality. In this capacity, the owner does not supervise but rather gives the contractor a set of expectations to meet, inspects and tests the work as it is developing (and when completed), and either accepts or rejects the results. *Performance monitoring* is a term used to describe this more passive, less direct involvement in the activities of the contractor. Specific techniques employed by owners to assure themselves that contractors are producing work in accordance with the contract requirements include:

Inspection
Testing
Review and/or approval of contractor submittals
Guarantees, warranties
Independent certification
Quality assurance criteria
Acceptance requirements

These are all elements of performance monitoring, not necessarily contract management, and the distinction is significant.

Whereas performance monitoring and its associated corrective action answer the question, Are we getting what we are paying for? contract management addresses a variety of ancillary questions. These range from, Is the contractor maintaining adequate insurance coverage? Are changes in the work being priced correctly and paid for expeditiously? and, Are proper costs records being maintained? to Are claims or disputes being handled on a timely and fair basis? Figure 2 summarizes these contract responsibilities of the owner's representatives and the contractor.

Who Manages Contracts?

Although both performance monitoring and contract administration may be undertaken simultaneously by the same person or group, each is a distinct function requiring special talents and training. Even though they are closely associated by their very nature, a conflict of interest may develop when performance monitoring and contract administration are handled by the same person(s). For example, a construction engineer representing the owner may find it difficult to see that a certain section of work be completed by a contractor in order to meet schedule commitments, even though the work is not within the scope of the existing contract. In his or her zeal to see this accomplished, the engineer may be inclined to forgo the necessary contractual modifications (including cost quotations, acceptance by the owner, and written orders to proceed) that are contract administration functions.

Traditionally, the duties of formation and administration of construction contracts have been assumed by the project owner or, in many cases, the party responsible for preparation of the contract documents—such as the architect or engineer.

On large projects with many prime contractors, all working under separate agreements (sometimes as many as 100), a case can be made for centraliza-

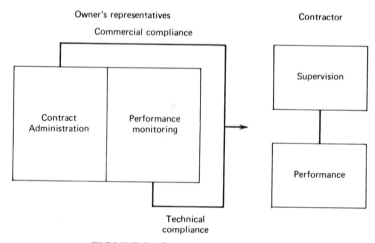

FIGURE 2. Contract responsibilities.

tion of contract management, leaving performance monitoring to several well-versed in the particular trade or construction process they are monitoring. Just as uniformity among all of the separate construction contract documents is a primary goal of those preparing the documents, so should uniformity and consistency be attempted in the management of contracts. Allocating all contract management duties to one person or group of people, especially on multiple-prime-contract jobs, can make good sense.

GENERAL RECOMMENDATIONS

A review of the topics in this book gives an indication of both the duties and identity of a professional contract manager.

Emphasis is placed on proper documentation and meticulous record keeping. Expect rigid adherence to provisions of the contract yet be flexible and open-minded enough to handle unforeseen conditions or unanticipated changes, which are inherent to the construction process.

The ability to get along with people from many social and economic backgrounds and representing many diverse interests also is essential. Coupled with a sizable amount of tact, it will preserve both your ability to perform and your sanity.

Ground Rules

Once a contract manager or one who will perform these duties has been chosen, his or her first mission is to establish ground rules covering conduct and relations with others during the period of the assignment. Basic ground rules should cover the areas of responsibility assigned to the contract manager (and—just as important—those *not* assigned), working and reporting relationships vis-à-vis other members of the organization (project management, the construction manager, estimator, accountant, engineers, and so on), and his or her authority (or limits of authority) to initiate or approve action.

It is imperative to have a thorough knowledge of all contracts, to establish oneself as the expert when it comes to the content of the contract documents.

One final point: contract managers, due to the very nature of their work, can easily forget or overlook the need for review and advice from qualified legal counsel. It is usually impractical to insist that all these duties be performed only by attorneys, and it is paramount that those who are not qualified to practice law refrain from doing so. On the other hand, the sole reliance on legal counsel to prepare technically and commercially sophisticated contract documents is just as grievous an error. Lawyers are generally unfamiliar or unconcerned with the operational ramifications of contract wording. Legal review usually limits itself to questions of liability and enforceability. Project and contract managers, by virtue of their experience, education, and vested interest in the day-to-day problems that contract wording, or omissions, may create, have a different perspective. For example, although a lawyer may point out the proper "hold harmless" or insurance requirements needed, he or she should not be relied upon to ensure that the contract ties progress payments to specific, quantifiable milestones or that as-built drawings be approved as a prerequisite to final payment. Once again, as is so often the case, common sense and good judgment should prevail.

A System Is Essential

Without a logical system by which to operate, the contract manager will be lost in no time. His or her job is to bring order to the contracting process. Without adhering to a minimum amount of regimentation, he or she can fast become a passive member of the construction team reacting to events rather than foreseeing or controlling them.

A common way of establishing a system is through project procedures and working instructions, such as those contained in this book. This up-front planning will save countless subsequent arguments concerning who does what to whom. A filing system should be established, manual or automated, as should be the format and forms of paperwork you will handle. Simplicity and uniformity should rule when developing standard documents. Checklists, logs, and standard forms, such as those included in this book as exhibits, are extremely helpful in bringing order to operations. Not only will they simplify work by streamlining it as much as possible, but they will also provide the basis for the airtight documentation required for claims management and for concise and consistent record keeping—a blessing when audit time arrives.

All of the above should not be construed as making the contract manager a glorified clerk. It is indisputable, however, that a certain amount of "everything in its place and a place for everything" thinking is not only tolerable but also desirable. Discipline and flexibility are both needed. Striking the right balance between the two is the greatest challenge of contract management.

CHAPTER 2

ORGANIZATIONAL AND CONTRACTING STRATEGIES

Once an owner decides to embark on a construction project, the major decision it must make is how best to organize the effort and delegate work to outsiders. This decision affects not only the management of the project but also the number, scope of work, and responsibilities assigned to internal and external organizations involved. In this chapter we describe major organizational alternatives, considerations for their selection, and the resulting impacts such a selection has on contracting processes.

CONTRACTUAL IMPACTS OF THE ORGANIZATIONAL APPROACH

Because the purpose of this chapter is to indicate the impact an owner's approach will have on subsequent contract management activities, rather than the merits of each approach alone, it is important to understand the far-reaching ramifications of such a decision. First, the chosen organizational approach determines the number of contractors that will be involved. In this manner, it defines the extent and limits of contracting—dictating the contract formation and administration efforts required.

The organizational approach also sets the framework for project-related authority and general responsibility assignments. The construction contracts further expand and define this authority and responsibility among the organizations involved. They follow the pattern of the organizational approach by assigning roles and responsibilities to those who will be operating under contractual relationships. In addition, because an organizational approach implies a general scope of work for each participant, the construction contracts must clarify and communicate the details of that scope. Construction contracts are the tools by which an owner implements the specifics of its organizational strategy.

ORGANIZATIONAL IMPACTS OF THE CONSTRUCTION CONTRACTS

Not only does the project's organizational approach affect the contracts, but the converse is also true. That is, the contracts themselves impact the organizational approach. For example, the contracts establish project-related authority, they distribute scope of work among project participants, and they specify the performance standards that must be met by those under contract with the owner. They also define the consideration—that is, money—each participant is to receive for project-specific work and identify the technical and commercial controls for the contract and the project as a whole.

When discussing contractual controls, therefore, we must realize the close association between the contracts themselves, the various contractual relationships they establish and enforce, and the project organizational setting in which they apply.

CONSTRUCTION'S PROJECT ORIENTATION

The construction of a major facility differs markedly from other large efforts requiring a great deal of time, money, and human resources. The most important difference is that construction is "project oriented"—that is, a unique collection of people, equipment, and material is brought together to create a finished product at a unique location. Although most of the tasks required to construct a facility have been performed many times before, the collection of tasks performed by the particular people, at a particular time, at a unique location, and for the particular needs of an owner constitute a one-time, unique project.

Contrasted with line consumer manufacturing—the mass production of automobiles, for example—where a relatively constant set of people, equipment, and activities produce a relatively constant product over and over again, construction of a facility occurs only once. In most cases, the collection of resources—people, material, equipment, and so on—needed for a project has never been brought together before and will never be assembled in the same manner for the same purpose again. The only common point of reference for all project participants is, therefore, the project itself.

If one looks at the example of automobile production from an organizational perspective, it is fair to assume that an operating plant producing these items does so under an established, ongoing organizational structure. This structure exists under the authority of one company and uses established lines of reporting and communication. Each member of that organization, presumably, understands his or her role and performs activities in order to meet individual long-range goals and those of the organization.

These features are not evident with a construction project. First of all, the organization involved is not established and operational. It must be created. Lines of communication need to be established, and there will be many authorities involved—sometimes operating with conflicting priorities. Because the project represents a temporary, one-time effort, short-term goals and relationships predominate over the long-term variety.

Functional contrasts may be drawn between the automobile production example and construction's project orientation. Line manufacturing requires the repeated production of identical or similar products using a series of tasks performed over and over again. Construction, on the other hand, involves unique, one-of-a-kind products, produced through a unique set of tasks. Production steps taken with mass assembly are linear—that is, they are performed in a structured sequence. This is not the case with construction. It involves steps that take place collaterally, or in parallel. Transient production facilities and methods are used in construction, such as mobile cranes, specialized rigging arrangements, and portable welding machines, whereas line manufacturing's facilities are fixed.

A primary difference involves the owner of the finished product in the production process itself. For line manufacturing, the owners (in this case, the potential buyers) have no direct impact on the manufacture of the product—they are not even identified until the products are completed. Owners engaged in construction efforts, on the other hand, are vitally interested in each step of the effort as it progresses. They choose the project participants, delegate responsibility to each, and manage or monitor the project throughout its life.

Many of these distinctions may seem fairly self-evident, but each affects every aspect of the construction industry and has shaped the way its business is conducted. Each points out the importance of the contracts themselves, as well as the demands of contract management. For it is the contracts that establish the roles and relationships that must be created, define the production organization, and provide incentives and penalties in lieu of long-term relationships and goals that exist elsewhere. Considering the project-oriented nature of construction, then, the contracts under which it is performed and the management of those contracts represent major factors for success or failure.

TYPICAL ORGANIZATIONAL AND CONTRACTING STRATEGIES

There are a myriad of ways by which an owner may organize and manage a construction project. Each has proven successful when properly planned and implemented. And each has established the framework for failure when used by those unfamiliar with its features, strengths, and concurrent obligations—

or when inadequately implemented. In describing the features, benefits, and disadvantages of each, two important facts must be kept in consideration:

1. No approach is inherently superior in all cases. Each has been used favorably and unfavorably in a number of cases. Although proponents of various approaches have attempted to justify the superiority of each, there remains no consensus in the industry that would lead one to choose any approach over the other without considering other owner- or project-specific factors.
2. Combinations of approaches are fairly common. There is no reason to ignore possible "hybrid" approaches incorporating aspects of several described here. In fact, many projects employ different aspects of each approach during different phases or for different portions of the project. Indeed, seldom is one "pure" organizational approach chosen and maintained throughout a project's life, particularly when the project extends over several months or years. Quite often, variations are adapted to meet specific or changing objectives, economic conditions, and management philosophies.

Five general approaches are described in the following sections. They are represented by generic organization charts in Figure 3 in an attempt to point out salient features and differences among them. In addition, an example "hybrid" approach is also illustrated to indicate only one of many thousands of combinations available.

Design–Build

This is an extreme example of the project owner delegating responsibility to an outside party. As we shall see later, it is the opposite of the final approach depicted (force account), where the owner assumes the dominant, if not total, responsibility for the project's completion.

With the design–build approach (also called a turnkey contract) the owner selects one firm to both design and construct the facility. It gives the design–build contractor a set of criteria it expects the completed project to meet, such as operating and performance characteristics, cost, site location, aesthetic appeal, date required, and general preferences—for example, similarity to an existing facility, use of local materials, architectural style, and brands of equipment preferred. Within these general parameters, the contractor must design the project and build it as well. In some cases, the contractor agrees to operate the facility over a period of time, either selling the products to the owner at a predetermined price or leasing the facility from the owner at a prescribed fee.

Often, the design–build contractor chooses to subcontract portions of the work to others. The design–build concept, however, still remains because the

FIGURE 3. Organizational approaches.

...tract between the contractor and the owner—not the degree of ...ocontracting—is the distinguishing characteristic of this approach.

The assignment of responsibility under a design–build approach is straightforward. Except for certain owner duties, such as payment for work performed, review and approval of intermediate and final products, and general decision-making obligations, the majority of project responsibility lies with the design–build contractor, who is responsible for planning, conducting, and coordinating all tasks necessary to produce the finished facility. Even with this concept, however, there are many responsibilities that are assumed by the owner unless specifically assigned to the contractor or others by the contract. These must be carefully considered and addressed by the owner prior to engaging a design–build contractor. Among these are licensing, real estate acquisition, permits, design approval, accounting and financial control, reporting to outside agencies, safety, and security.

A major commercial risk associated with the design–build concept concerns the lack of management flexibility it affords the project owner. By selecting one company to design and to construct the facility, the project's success becomes directly dependent upon the financial stability, management controls, and operational effectiveness of that company. Should any prove unsatisfactory, it is difficult to remove the contractor for the project without severe cost, schedule, and technical impacts. These impacts become greater as the project progresses. As will be discussed further in the next chapter, the design–build approach is rarely contracted in a firm-fixed-price manner. Because of the inherent uncertainty in scope of work (that is, the design has yet to commence) cost-reimbursable pricing structures are most often employed when contracting for design–build services.

From an owner's standpoint, the most critical performance control is the selection of the design–build contractor. Since detailed controls over cost, schedule, and technical performance will most often be implemented by the contractor, the owner must satisfy itself that these controls are adequate before the contract is signed. This requires extensive bidder qualification and a detailed examination of specific contractor controls during the selection process. Requirements for contractor controls should be specified in detail in the contract documents, and the owner must continuously monitor their use. It delegates detailed control responsibility to the design–build contractor, so it must assure itself that the contractor has the intent, personnel, and systems to exercise such control. The owner's duty then reverts to surveillance of the contractor's controls throughout the project duration.

As we shall see later, the degree to which the construction project must be defined or specified prior to awarding a design–build contract is slight. This is an advantage to its use. Additionally, there is only one contract to form and administer from the owner's point of view. Impact on the owner's staffing and management requirement is minimized with this approach. But commercial risks are not entirely avoided.

General Contractor

This is the traditional contractual approach most often seen in smaller building construction projects, such as hospitals, churches, schools, and shopping centers, but used quite frequently for industrial projects as well. With this approach the owner contracts with an outside engineering and design firm (architectural–engineering firm, or A–E), which will design the facility, and also contracts separately with a large construction contractor to build it. Most large general contractors will, in turn, subcontract portions of the construction work to other smaller or more specialized contractors (subcontractors, or "subs").

The difference between this approach and design–build is that the design function is awarded separately from the build function. A third party, the A–E, is brought under contract with the owner rather than as a part of or subcontract to the constructor's organization. The owner is now dealing with two distinct organizations under contract. Both—A–E and general contractor—must interface with each other excessively in this arrangement. As opposed to the design–build situation, the awarding of the design contract precedes award of a contract for construction. The products of the design effort (drawings, specifications, and so on) are needed to allow general contractors to determine the scope, cost, and schedule of their work. Ideally, all of the design is completed prior to the bidding of a construction contract. The work is well defined, a schedule can be readily determined, and a price quotation can be made by the bidding contractors with some degree of confidence. In actuality, this is seldom the case.

The general contractor approach provides for a classic division of project responsibility between engineering and construction efforts. The owner is not typically involved in subsequent delegation of responsibility within either of these contracted efforts. And coordination of work performed by any subcontractors to either the A–E or general contractor is the responsibility of these respective companies.

Two major challenges confront owners contemplating this approach:

1. Identifying and assigning responsibility for functions that, although required for all projects, are not automatically associated with either design or construction. This includes such functions as licensing, permits, procurement, and vendor expediting, among others.
2. Coordinating and communicating between the design and construction companies throughout the project life.

Some owners using the general contractor approach delegate much responsibility for non-design-specific functions to the A–E firm. Depending on the extent of this delegation, the A–E may also perform other functions as a representative or agent of the owner. Some functions in this category are procurement and procurement-related activities, project administration, contract administration, and inspection and testing. To the extent that this occurs, the owner may greatly reduce its direct project involvement. In any case, however, the owner must be prepared to coordinate, and sometimes mediate between these two contractors (design and construction) before selecting this approach.

The "one-contractor" risks inherent in the design–build concept also apply to the general contractor alternative, but to a lesser extent because engineering and construction are performed by separate companies. Since design is rarely performed by more than one company, this effect is more meaningful in the construction context. Reliance on a cost-reimbursable pricing method for the construction portion of the work is not as severe as with the design–build concept. In cases where significant design has been performed before a general construction contract is awarded, the benefits of fixed pricing may be pursued. Two factors contribute to the fact that firm fixed pricing is seldom encountered, however, for the entire scope of construction:

1. Design is seldom totally or substantially complete at time of construction contract award.
2. Inflation, escalation of material and labor prices, and changes in the work invariably occur during the length of time required to perform all project construction work. These changes, as we shall see in the next chapter, are normally reflected in unit price or cost-reimbursable pricing clauses.

The added burden of coordination placed on the owner also increases demands on owner staffing and system requirements. When owners cannot meet the challenge of this increased involvement, the risk of an unsatisfactory performance increases.

Because coordination of these two major efforts is typically the responsibility of the owner, owner controls must be put in place as soon as an A–E is selected. These include provisions for sequencing of design work to meet the needs of construction, timely and adequate review of design products (drawings and specifications), schedule management, and communication control. When owners accept direct responsibility for functions that span design and construction, such as procurement, scheduling, and project reporting, owner controls over these functions must be implemented.

Few Primes

The few-primes approach is not as distinct as others described in this chapter. It can best be thought of as representing the "middle ground" between the extremes of design–build and force account. It occurs when the owner awards more contracts for construction than are used for the general contractor approach—that is, more than one—yet fewer than those encountered by owners under a multiple-prime strategy. A typical few-primes situation is where separate prime contracts (as opposed to subcontracts) are awarded between the owner and, say, five contractors. In most cases the subdivision of work among these prime contractors is along product or discipline lines, such as civil, mechanical, and electrical. There may be one contractor designated as the "general," but all other contractors are not necessarily subcontracted to it.

For example, the owner may award separate contracts for inspection and testing services, painting, and electrical work, with everything else being performed by the general contractor or subcontracted by him. Indeed, the owner may also elect to buy certain equipment and material from suppliers and furnish these to contractors for installation during the construction process as well. The degree to which the owner becomes directly involved in the selection of contractors and suppliers determines the associated responsibility it must assume for their coordination. Under the umbrella term of *few primes*, the owner's involvement may vary greatly.

As with previous alternatives, the design function is performed internally or given to one outside firm. The construction effort must be thoroughly planned, packaged into distinct scopes of work, and awarded so that each item of work is assigned to one and only one prime contractor. This effort becomes more demanding and risky as the number of contracts increases and the allocation of work scope becomes less traditional.

Under this approach the owner assumes many detailed coordination, administration, and scheduling responsibilities normally performed by the general contractor under the approach using that term. The owner must insert itself not only among the few construction contractors but also between each contractor and the A–E firm. In some cases owners engage the services of an outside "project management" or "construction management" firm to discharge these coordination duties. This is not a requirement of the few-prime approach, but it often eases the detailed demands on the owner's staff and systems.

The risk of investing all construction work with one contractor is reduced with the few-primes approach to the extent that more contractors are involved with the construction effort. Should any one contractor experience stability or performance problems, it may be more easily terminated or replaced than if it were responsible for all construction work.

Cost risk may be reduced through the selective use of firm-fixed-price contracts, sequenced for award after complete design information is available. Shorter construction durations for each contractor also reduce reliance on cost-reimbursable pricing. The greatest element of risk to a project owner operating with this approach involves the use of smaller (perhaps less stable) contractors whose performance is greatly dependent upon owner management, coordination, and control capabilities. The lack of a general contractor to schedule, sequence, and coordinate the construction work, including the administration of construction disputes, requires the owner to increase its project involvement. To the extent that an owner cannot meet these obligations, its risks increase dramatically.

As with the multiple-prime approach, this approach depends on:

1. Discrete, well-defined subdivision of work among the associated contractors
2. Thorough qualification of potential contractors
3. Selective control of cost, schedule, and technical aspects of each contract based on performance risks
4. Proper level of owner involvement to manage and control each contractor's work

Multiple Primes

When there is no dominant or "general" contractor, and most of the outside organizations are performing under separate, direct contracts with the owner, a multiple-prime situation exists. In a sense, the owner has replaced the general contractor in its role of subdividing the work and awarding individual subcontracts (in this case, *contracts*). Needless to say, this increases the owner's participation one level greater than the previous approach. The owner (or its agent) becomes the ultimate coordinator of all contractors on the site. Often, the multiple-prime approach is called *construction management*. It also requires the issuance and administration of more contracts than do all other approaches. The main disadvantages of a multiple-prime approach are easily apparent: extreme demands on the owner (and A–E) for staffing, responsiveness, and organizational strength. Many owners fail to realize this and undertake this approach without preparing for the increased responsibility, additional staffing, and assumption of risk that it entails.

The multiple-prime approach does have its advantages though. For one, the package of work into many small contracts makes the use of smaller local contractors possible. In addition, a type of "portfolio effect" is achieved. The owner is not captive or unduly dependent on one large contractor once construction is under way. It has the flexibility of awarding each package on its own merits as far as pricing structure is concerned. Should, for example, one segment of the work be undefined, the owner can hold off on its award until it is. Once construction begins and several contractors are mobilized, the possibility of their being awarded later portions of the work, should they perform well, is a valuable incentive.

A condition called "fast tracking" (either planned or unplanned) exists when portions of the design, or construction "packages" are released from the A–E to the contractor(s) in advance of other design products, which are in preliminary stages of development and are not immediately required. This is advantageous in getting the construction started as early as possible—for example, foundations can be excavated and poured before electrical equipment and wiring has been designed—but it intensifies scheduling and coordination demands.

Because the multiple-prime approach lends itself to the staggering of small design and construction efforts, it is commonly seen when fast tracking is used. Fast tracking has been heralded as a viable time-saving technique, with overall project schedule compression as its advantage. It requires that design be several steps ahead of construction. However, at times design falls behind or rushes out incomplete and unsatisfactory products. Therefore, rework and changes to the construction contracts are common. Potential rewards are high, but so is the risk involved.

Force Account

Force account is the owner's ultimate step toward project involvement and responsibility. It is the classic "do it yourself" approach. In its extreme case, the owner designs and/or builds the entire project, using its own personnel and equipment exclusively. As in all approaches described here, the owner has dedicated personnel assigned to the project. But with the others, the owner's people perform management, supervisory, or monitoring functions. With force account, the owner also has the actual craft workers (carpenters, ironworkers, bricklayers) on its payroll. The workers (forces) are to its "account." Usually, even with a dedicated force account owner, some specialty work is contracted to outsiders.

A variation of the force account approach occurs when, rather than placing construction craft workers on its payroll, the owner hires a "labor broker" to furnish the workers. With the labor broker method, the broker differs from a contractor in that it assumes no risk for contract completion and is not responsible for the work itself. This is left to the owner, with the labor broker simply furnishing workers on an as-required basis. For this service, the labor broker is paid a fee, which is generally a percentage of the total wages paid to the workers provided. The workers are employed by the broker but used and directed by the owner.

Needless to say, a force account approach places extreme demands on the owner. That is why most owners—except for those with extensive long-term construction programs—avoid this strategy. Even continual builders, however, cite the following reasons for passing up the opportunity to go into the construction business:

1. Possible adverse reaction of outsiders (contractor organizations, stockholders, and so on)
2. Human resource limitations
3. Personnel recruiting, training, and retention costs

4. Need for large equipment and supply inventories
5. Construction labor relations difficulties
6. Increased liability for construction-related tasks such as transportation, logistics, security, and safety

SUMMARY OF MAJOR FEATURES

In reviewing the five major organizational approaches described in this chapter, several conclusions may be drawn. These are summarized in Figure 4 and described in the following paragraphs. Only by thoroughly understanding the features and benefits of each approach can you: (1) make a rational selection of any one approach over the others; and (2) successfully implement the choice they have made.

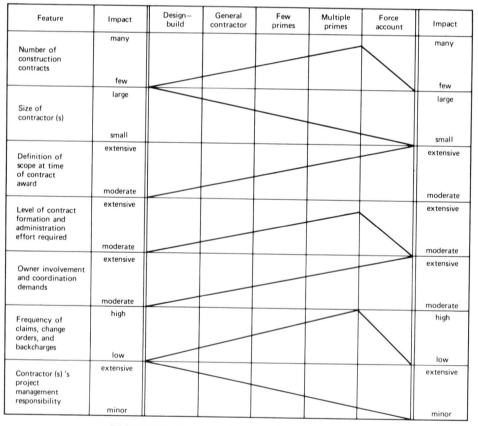

FIGURE 4. Impacts of contracting approaches.

1. *Level of Owner Involvement Required.* The major effect of an organizational approach is to dictate commensurate owner involvement in project-specific activities. And depending upon the subsequent assignment of responsibilities within the boundaries of any given approach, the owner's involvement varies greatly.

2. *Contract Pricing Methods.* Although described in more detail in the next chapter, pricing alternatives are dependent to a great extent upon the organizational setting of a construction project. Absent other influencing factors, such as the competitive nature of the contractor marketplace, firmness of design information, and general economic uncertainty, the ability to secure firm-fixed-price contracts increases as each contractual performance period and scope of work decreases.

3. *Single Contractor Risk.* As the number of outside companies involved in a project increases, the vesting of risk in any one decreases. Negative impacts resulting from failure, termination, or inadequate performance by any one contractor are reduced when the total project scope is distributed among many outside organizations. The benefits of this "portfolio effect" on risk must be weighed against increased management and control demands.

4. *Contractor Control Capabilities.* As the relative scope of work and size of a company increases, its internal management and control capabilities typically improve. In this regard, larger design–build contractors usually maintain more extensive management and project control capabilities in comparison to smaller, specialized contractors who typically work on restricted scope, within the environment of controls provided by others.

5. *Contract Formation and Administration Requirements.* Two major factors influence the scope of contract management duties and the difficulty with which they are performed. These are the organizational approach selected by the owner and the pricing structures used for each contract. Absent pricing considerations, contract formation and administration efforts increase in direct proportion to the number of contracts used for a given project.

CHAPTER 3

CONTRACT PRICING ALTERNATIVES

There are an infinite number of ways to price and pay for contracted goods and services. In this chapter we explore the most commonly used methods from the viewpoint of the features, element of cost risk, suggested control methodology, required level of owner involvement, and applicability of each.

The most common ways to price contracts are shown in Figure 5. They are arranged in major groupings of those considered fixed-price techniques and the remainder in the cost-reimbursable category. As shown, they range from (1) firm fixed price or lump sum, where virtually all cost risk lies with the contractor, to (2) cost plus percentage of cost, where all the cost risk rests with the owner and, in fact, the contractor has a positive incentive to increase the cost of the work. Within these extremes, there are almost unlimited possibilities and combinations. Because of the range of these alternatives, it is sometimes impossible to categorize clearly a given contract as either "fixed price" or "cost reimbursable," as it may encompass aspects of both in an attempt to incorporate reasonable incentives and risks from the standpoint of both buyer and seller.

PRICING TERMS

In our discussion of various pricing alternatives, several terms will be used. Each has a precise definition, and only by understanding the meaning of each can one expect to understand, or unravel, the sometimes confusing distinctions encountered with pricing alternatives used in the construction industry today. Basic terms used with regard to each alternative are:

1. *Cost.* This is defined as the seller's (contractor's) actual cost; that is, the price it pays for labor, equipment, material, and so on.

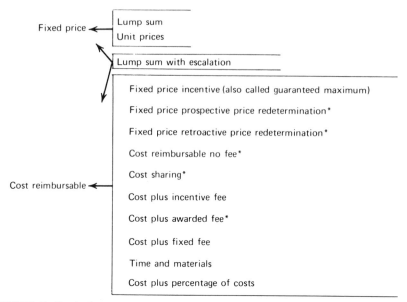

FIGURE 5. Typical contract pricing alternatives. (Pricing methods marked * are seldom encountered in conjunction with the construction of major commercial or industrial projects.)

2. *Fee.* This represents an additional amount to be paid to the contractor over and above its incurred costs. It usually, but does not always, represents payment for his overhead, profit, and expenses not included in the above-defined cost. As will be seen in the descriptions of some pricing alternatives, this fee may be positive (contractor gain) or negative (contractor loss).

3. *Price.* Price represents the total amount paid by the owner to the contractor. In most cases, it represents the sum of the contractor's cost and fee. The contract price can also be described as the owner's "cost" for having the contracted work performed.

CONSTRUCTION CONTRACTS VERSUS PURCHASES

The term *contract* refers to those arrangements in which job-site labor is involved in the seller's scope of performance. This distinguishes *contracts* from *purchases,* where no job-site labor is involved on the part of the equipment or materials vendor. An example of a purchase would be the furnishing of pumps, motors, or electrical cable to the owner or to a contractor by a supplier. Of course, job-site labor would be required to install these items, and the installation would be covered by a separate construction contract of some kind.

As mentioned earlier, construction contracts can include the furnishing of labor and its tools alone, such as foundation excavation, or labor and materials or equipment. The latter contracts are called furnish-and-erect or furnish-and-install contracts, and an example is a contract for furnishing and erection of structural steel. The pricing alternatives described in this chapter can be used on conjunction with either furnish-and-erect or labor-only contracts. If no site labor is involved, we call the seller a *vendor* and typically issue a purchase order, not a construction contract.

CONSIDERATIONS FOR SELECTION OF PRICING METHODS

Several factors influence the choice of a pricing alternative in any given situation. They affect the contractors' ability to provide bids or firm estimates as well as the ease with which progress can be measured and paid for during the performance period. Before deciding on any pricing method consider these suggestions.

1. *No single pricing method is superior in all cases.* Although a general consensus among owners favors fixed-price types of contracts over the cost-reimbursable variety, the latter may prove far more cost-effective given certain conditions. Each pricing method may be totally appropriate given certain owner objectives, scope of work, bidding information and bid circumstances, and market conditions and yet wholly inappropriate given others. Contracting strategy (including pricing method) should be based on a thorough analysis of these factors on a case-by-case basis. In general, every effort should be made on the part of an owner to plan and control contracting efforts to facilitate fixed-price buying. However, recognize that cost-reimbursable contracting has benefits when appropriately used, carefully defined, and closely monitored. General relationships among pricing methods, owner cost risk, and cost monitoring are shown in Figure 6.

2. *Combinations of pricing methods are common.* It is not uncommon for owners to use several different pricing methods within the framework of a construction project. In addition, many contracts contain provisions for a combination of pricing methods, each applying to different scopes of work or under differing contractual or project conditions. An example would be a contract for soil and rock excavation. A lump-sum (fixed-price) amount may be agreed to for the major foundations, whereas unit prices are employed for additions or deletions (adds and deducts) to the anticipated volumes of soil and rock. Cost-reimbursable terms may also be employed for additional work not included within the scope covered by the lump-sum amount, such as clearing and grading of access roads yet to be designed. If the performance period is long, escalation terms applying to increases in material and labor costs may also be applied. Not only can different pricing methods be em-

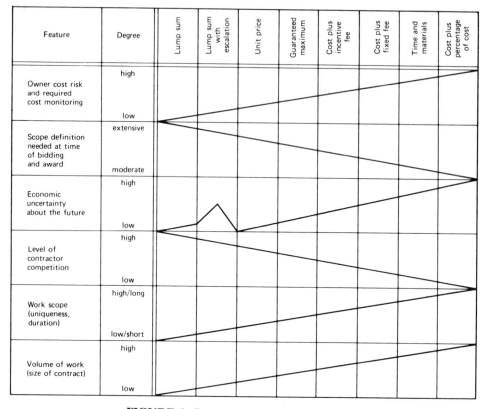

FIGURE 6. Features of pricing alternatives.

ployed within the boundaries of a single contract, in many cases that is more sensible than strict adherence to a single method.

3. *Performance incentives and escalation provisions may be added to each pricing method.* Regardless of the pricing base on which goods or services are bought, additional pricing criteria are often included. These include bonuses and/or penalties based on the cost, timeliness, or quality of the goods or services provided. Escalation provisions, designed to protect the contractor in the event of increased costs due to inflation or other factors beyond its control, can also be included. Performance incentives (both positive and negative) and escalation provisions are described in more detail later in this chapter.

4. *Pricing methods must be chosen deliberately, defined properly in contractual agreements, and carefully monitored to achieve their objective.* Selecting the best pricing method (or combination of methods) is only the first step in a series required for cost and performance control. Precise definition of the pricing basis must be included in the request for proposal (RFP) and contract documents. Methods for measuring, accepting, and paying for work must be thoroughly specified and understood by all affected parties. And careful monitoring of performance, progress, contractual compliance, and

cost data must be ensured. As described throughout this chapter, the degree to which these activities must be performed varies directly with the pricing method chosen.

5. *Owners are not always free to select the pricing method.* Aside from the desires of the owner, there are a number of outside factors or influences that often affect the selection of pricing methods. These include:

Status of Scope of Work and Performance Information at Time of Bid and Award. The less defined either the scope of work or performance conditions, the less chance an owner has of employing fixed-price methods. This relationship is shown in Figure 6.

Expected Price Fluctuations or Economic Uncertainty. The more uncertain contractors (and their subcontractors and suppliers) are about future factors that may influence their performance and costs, the more reluctant they are to agree to fixed-price terms (see Figure 6).

Scope of Work and Performance Duration. Longer performance periods (time needed by the contractor to buy, fabricate, deliver, install, or otherwise perform its contractual obligations) and greater scopes of work generally restrict the use of fixed-price methods. This is due primarily to uncertainty or risk of adverse economic trends, such as inflation, or renewal of labor contracts, over long periods of time. Shorter performance periods provide contractors more confidence that factors adversely influencing their profitability will have less effect (see Figure 6).

Available Sources of Supply (Competition). The more acceptable vendors or contractors available, the more freedom the owner has to select pricing methods and contractual terms. Conversely, a small number of acceptable bidders usually reduces the owner's bargaining power, including the selection of pricing alternatives. In other words, the environment at the time of contract bidding or negotiation may be either a "buyer's market" or a "seller's market." When market factors encourage competition among sellers, the owner's ability to dictate pricing methods is enhanced. This relationship is depicted in Figure 6.

Uniqueness of Scope. Fixed-price agreements are more applicable to performance that is easily understood, priced, and conducted. Standard or relatively simple tasks, or those that have been accomplished many times by contractors, lend themselves to fixed pricing. Unique, advanced, unproven technology or exotic practices reduce the applicability of fixed-price terms. Under unique or unproven conditions, cost-reimbursable methods are generally used. This relationship is shown in Figure 6.

6. *Subtier pricing need not follow prime contract pricing.* Regardless of the pricing method(s) used in owner-contractor agreements, the contractor's agreements with suppliers and subcontractors may be different. It is true, however, that subtier agreements may influence a contractor's preferred pricing method within an owner contract. For example, contractors who can arrange only cost-reimbursable subcontract or supply quotations are gener-

ally reluctant to offer owners fixed-price bids. By the same token, owners using cost-reimbursable pricing terms may have some say in future subtier pricing decisions. Although there are cases where subtier and prime contract pricing methods affect each other, the relationships are not direct; nor are they mandatory.

7. *Services are generally more difficult to price than goods.* Because of the inherent variables associated with providing services, they are generally more difficult to price than goods alone. This is particularly true when comparing pricing methods prevalent in the purchasing of material and equipment—concrete, structural steel, cable, electric motors, and so on—and those employed for service or construction contracts—design services, furnish and install electrical instruments, piping installation, and so on. Because the manufacture and delivery of goods typically takes place under controlled conditions (a factory rather than a construction site) and has generally been performed before (mass assembly versus one-of-a-kind installation), the purchase of these goods is more conducive to fixed-price methods. Design or construction services, on the other hand, are often influenced by (1) greater owner involvement and owner-directed change; (2) environmental factors, such as weather, access, labor availability, and restrictions in work space and methods; (3) transient, temporary work forces; (4) mobile production facilities, such as cranes, welding machines, and earthmoving equipment; and (5) uniqueness of scope. For these reasons, fixed-price methods are more prevalent when buying material and equipment (purchases) than when buying services (A–E, construction contractors). However, the presence of other previously described factors, such as economic uncertainty, and uniqueness of scope, also reduces the applicability of fixed-price methods within the context of purchase-only agreements.

8. *Creative work is more difficult to price than goods or services.* When you are buying one-of-a-kind objects or services, pricing guidelines and competitive discipline are not very helpful. Take the work of an artist or sculptor, for example, or someone who is painting a mural in the lobby of your corporate headquarters. Here there is little competition, you can not use unit prices, and cost-plus-a-fee becomes a little ludicrous! Most of this type of work is priced on a fixed basis, and the time-honored "what the market will stand" approach is used for the quotation. Owner controls rely on (1) selection of the source (artist, designer), (2) specifying what you want well in advance and in detail, and (3) periodic review of the work as it progresses.

CONTRACTOR PERSPECTIVE

Pricing methods represent a clear example of differing owner and contractor perspectives, and for very real and understandable reasons, for the basic concept of pricing is that it represents an attempt to share risks between the two. In the case of cost risk, for example, virtually every sta gestion, or conclusion drawn in this chapter should be reversed contractor's perspective. Where fixed-price contracting genera an owner the most protection from cost risk, it simultaneously exposes the contractor to that same risk. The same relationship holds for owner involvement in monitoring and managing the work. Cost-reimbursable pricing demands close owner scrutiny of contractor efficiency and purchasing practices. Most contractors scrutinize these same practices more closely when they are operating under fixed-price terms—because cost savings under these conditions accrue solely to them.

Because contractors are in the business to make a profit, the concept of cost risk can also be viewed as "profit opportunity." Given a choice, many contractors prefer a well-defined and comfortably priced lump-sum contract over a cost-reimbursable one in which percentage markups for profit and overhead are small. If a contractor feels the risks of overrun are minor with the lump-sum arrangement, it stands to reap more profit than with the cost-reimbursable contract. When the following pricing alternatives are reviewed, note that although they are described from the owner's perspective, there is a lot to be learned by considering the contractor's point of view.

PRICING ALTERNATIVES

Each principal pricing method is described in the following subsections, beginning with firm-fixed-price alternatives and continuing through those classified as cost reimbursable. Absent all other considerations (and there are many), the alternatives are listed in general order of preference from a buyer's (owner's) perspective.

Lump Sum

A definite and fixed price is agreed upon prior to contract award. This price remains firm for the life of the contract and is not subject to adjustment except for changes in scope of work or performance conditions and owner-ordered extras. Under lump-sum agreements, cost risk to the owner is minimal—given adequate bidding and performance controls. Major risks associated with lump-sum contracting include:

1. *The Possibility of Large "Contingency" Amounts Being Included in the Quoted or Contractual Price.* Knowing that there will be no changes in the contract price (except for changes to the work), contractors may artificially inflate the lump-sum amount to cover real or imagined cost risk to them.
2. *Changes to the Work.* Since these are not included in the lump-sum amount, they are priced separately and represent additional costs. They

usually cost more when added than if they were included in the original quote.

3. *Breach of Contract.* Should the contractor grossly underestimate costs and agree to an unreasonably low lump-sum price, it may fail to perform the work should his costs approach or exceed the fixed price. This may be unintentional—for example, the contractor loses so much money by performing the work that it goes bankrupt—or intentional—the contractor ceases work in order to concentrate on more profitable jobs. In either case, the owner suffers. Despite the fact that the contract may be legally enforceable, the owner faces schedule delays, quality of work problems, litigation costs, and other detrimental effects. Having a good price is not such a benefit if it is from a bad contractor.

4. *Reduction of Bidder Competition.* Strict adherence to fixed pricing when other methods are more appropriate could reduce the number of contractors wishing to undertake the work. As a result, the owner loses the beneficial effects of contractor competition.

There are three major ways to reduce or mitigate the cost risks associated with lump-sum pricing. They are:

1. Thorough bidder qualification procedures that subject potential contractors to strict financial, quality, and performance standards. This helps eliminate unstable contractors, wild bidders, and those without the financial resources to perform
2. Thorough definition of scope prior to bidding and award
3. Control of scope changes and extra work

Considering the above, lump-sum pricing is most applicable to short-duration, well-defined scopes of work. It is most commonly used for the purchase of off-the-shelf equipment or materials and short-duration, smaller construction contracts. It is less common for engineering or design services, consulting services, or long-term, large-scope contracting.

Unit Prices

Unit pricing represents a variation of the lump-sum method. Whereas lump-sum pricing involves one fixed price for all, or portions of, the work, unit pricing fixes only the price of a given unit or element of quantity. The total contract price is determined by multiplying unit prices by the quantity of items delivered, erected, or installed. With concrete placement, for example, the unit price may be $60 per cubic yard, installed. If 1,000 cubic yards were placed, the total contract price would be $60,000.

Owner risk with unit pricing includes many of those mentioned for lump-sum arrangements. In addition, unit pricing requires close monitoring and verification of the *quantity* of units actually furnished or installed. Keeping

track of how many were added, deducted, installed, or removed can be ᴜ
a chore.

Lump Sum With Escalation

Escalation provisions can be applied to both lump-sum and unit pricing. The
basic intention is to eliminate unwarranted price increases to cover contingency in the contractor's bid. Escalation provisions typically call for adjustment based on:

1. The actual experience of the particular contractor, such as how that
   contractor's cost changed from an anticipated or base amount, or
2. Published indexes of price changes for particular elements, such as
   labor or certain materials

The use of published indexes is generally more effective when determining
additional pricing due to escalation. This is mainly due to the ease with which
escalation amounts may be determined. However, when this pricing alternative is used, the following should be considered before award.

1. *Selection of Appropriate Indexes.* There are many available, some of
   which may be influenced by factors not experienced locally, and therefore inappropriate.
2. *Identification of That Portion of the Base Price to Which Escalation
   Will Apply.* Often only certain portions of scope are subject to escalation provisions. These should be carefully defined.
3. *Application of the Indexes to Progress Payments, Planned Expenditures, and Actual Expenditures.* This may prove a complex and difficult task.

Escalation provisions become quite common during periods of frequent
price fluctuation and for projects that require several years to construct. It is
much easier for owners to secure this type of pricing than to secure strict
lump-sum or unit prices during such times. They are very popular in countries
where inflation is high or unpredictable for certain materials that fluctuate in
price.

Cost risks to the owner are identical to those under lump-sum or unit
pricing, with the following exceptions:

1. Because escalation provisions protect contractors from potential cost
   increases, contingency amounts included in bid prices are generally
   reduced. This is a benefit to the owner.
2. Price increases as a result of general inflation and/or increases in the
   cost of specific goods or services are borne by the owner rather than by
   the contractor.

3. Escalation provisions may overcompensate the contractor beyond its actual experience (particularly when based on indexes).
4. Quite often escalation amounts are erroneously calculated, resulting in overpayment or payments that duplicate those made for the base (lump-sum or unit-priced) amount. Escalation provisions should therefore be drawn carefully, with consideration for potential misunderstanding or intentional manipulation.

To reduce these risks, owners should insist on clear definition and full enforcement of the items to be escalated, the method of determining escalation amounts, and the process used for payment of amounts due to escalation. Escalation provisions require more owner participation during the contract formation phase (determining escalation provisions) and during the performance period (verifying escalation charges) than required for nonescalated fixed-price agreements.

Fixed Price Incentive

Fixed-price-incentive pricing (also called *guaranteed maximum* or *guaranteed max*) allows the fee portion of the price to be adjusted (upward or downward) depending on certain factors that are under the control of the contractor. This adjustment is usually based upon actual costs incurred by the contractor during the work. The following factors are agreed upon and included in the terms of the contract:

1. A *target cost* for the work
2. A *target fee* for the work
3. A *target price* (target cost plus target fee)
4. A *ceiling price* that limits the buyer's responsibility for cost overruns, which is the *guaranteed max*
5. A *formula* for establishing the final contractor's fee and therefore the ultimate price to the owner, which is called the *sharing formula*

The relationships among these items for three typical contract applications are depicted in Figures 7, 8, and 9.

Fixed-price-incentive arrangements are used to provide a profit incentive to the contractor to reduce costs of its performance. They do this by providing a profit-sharing formula under which both the owner and contractor share in any reduction in costs and, conversely, in any burden of increased costs. A broad range of incentives are possible. This pricing alternative is increasingly coming into use in commercial contracts.

Major cost risks to the owner under such arrangements stem from two sources:

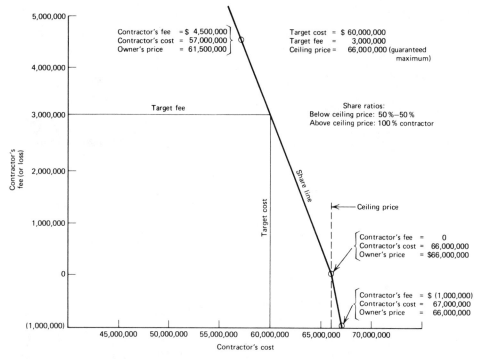

FIGURE 7. Fixed-price-incentive-fee graph 1.

1. Setting the guaranteed maximum (or ceiling price). Should this be set unduly high, there is little protection for the owner.
2. Selecting the cost-sharing formula. If the share of cost overruns is such that the owner assumes a major portion, there is less incentive for the contractor to maintain costs below the target cost.

These risk areas may be reduced by selecting the right target cost, target price, ceiling price, and formula for determining the final price (cost-sharing formula). Many owners, however, have difficulty monitoring the ceiling price as changes to the work occur. A major effort is usually required to determine the impact on the guaranteed max amount once scope changes, schedule delays, or contractor claims arise.

The following recommendations are given for those considering guaranteed maximum arrangements.

1. Perform an independent estimate of the likely cost of the work and use it to evaluate or verify contractor estimates of the target cost and ceiling price. Make every effort to determine a target cost and ceiling price that are reasonable and contain significant cost incentives.

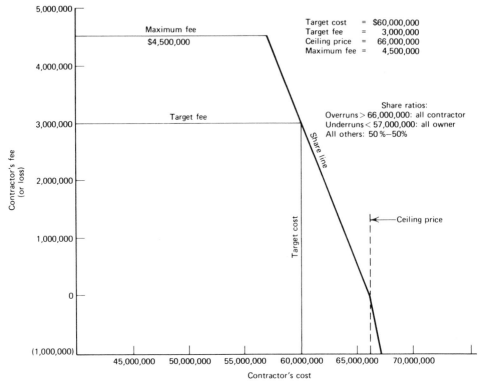

FIGURE 8. Fixed-price-incentive-fee graph 2.

2. Negotiate a meaningful and realistic target fee and sharing formula. A share formula that is heavily slanted in favor of the owner—for example, owner receives 95% of underruns, and contractor pays 95% of the overruns—has the effect of turning this contract into a virtual lump-sum arrangement. If the share ratios heavily favor the contractor—for example, contractor receives 95% of underruns, and owner pays 95% of overruns—the contractor has little incentive to reduce costs once the target cost has been reached.

3. Maintain quality standards. Contractors may reduce the quality of their work in order to cut costs and receive higher fees. Constant monitoring of quality should be maintained.

Although this pricing method requires less scope definition prior to award (as compared to a simple lump-sum arrangement), considerable owner involvement is necessary to establish the cost parameters required for an agreement. Setting the target fee and ceiling price and the formula for sharing cost overruns and underruns generally involve prolonged negotiations. Once the performance period begins, constant monitoring of changes and the determination of their impact on the target cost and ceiling price are necessary; in many cases, these also involve some form of negotiation. The potential for disputes and errors is high.

Guaranteed maximum pricing is applicable for goods and/or services that are difficult to price in a firm-fixed manner. It should be considered, however, only for those contracts where a reasonable estimate of probable costs may be established. Service contracts (such as with an A–E for engineering and design services) may also be based on this type of pricing structure. These contracts often use "target labor hours" rather than a target cost, with the fee adjusted depending on the number of labor hours used. Guaranteed maximum contracts are also common for construction, but usually apply to large general contracts rather than to smaller, specialty types of work.

Fixed Price Prospective Price Redetermination

A firm fixed price is established for a portion of the work, and provisions are made for price redetermination for future work of the same type. For example, the first portion of the contractor's scope may be priced as lump sum or unit price, with these prices agreed upon before work starts. However, once

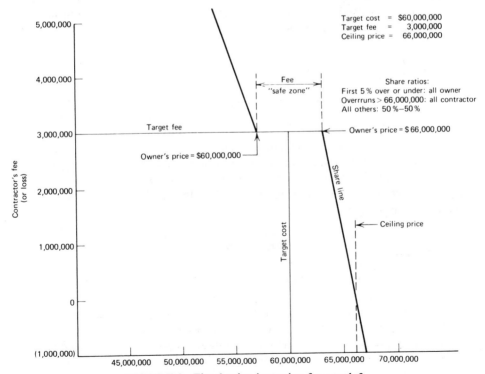

FIGURE 9. Fixed-price-incentive-fee graph 3.

that designated portion of scope is finished, remaining scope is priced under different conditions. Here is an example:

> The first 100 units of an item are priced at $15 each, with the provision that further unit prices will be determined at the time the units are ordered. Often, the future unit prices are limited by a ceiling. In this case, units beyond the first hundred may be guaranteed to cost no more than $20 each.

Risks associated with this type of pricing involve uncertainty at time of agreement regarding price and quantity of future items or work. In effect, this pricing requires the owner to enter into a series of fixed-price purchases without assurance of the fixed prices to be applied and includes the additional owner requirement to evaluate and negotiate each incremental order.

This pricing method is virtually unused for construction contracts. However, it is sometimes proposed by vendors for material and equipment purchases. Even for these cases, though, its use is rare.

Fixed-Price Retroactive Redetermination

A ceiling price is negotiated prior to starting work. After work is completed, a final price is negotiated, not to exceed the original ceiling price. This pricing method is rarely used. It has little or no application to construction projects and for that reason need not be discussed further.

Cost Reimbursable, No Fee

Under this arrangement the contractor is reimbursed for incurred costs only; no fee is paid. This alternative has little application in the world of commercial construction. When used, it is generally restricted to work performed for nonprofit organizations (universities, charitable foundations, and so on) or for research and development work in which the contractor gains other benefits by participating in the effort. A first-time construction or manufacturing effort where the contractor gains by testing its product or work methods may lend itself to such an agreement.

Cost Sharing

Both contractor and owner share the cost of performing the work. As with the previous alternative, this approach provides no fee for the contractor. In addition, the contractor is responsible for absorbing some of the costs of performing the work. It has little or no application to construction projects. Again, its use is generally restricted to nonprofit work, research projects, or perhaps to the case of construction of facilities that will be shared for the benefit of both owner and contractor when completed.

Cost Plus Incentive Fee

This method is similar to a fixed-price-incentive-fee (guaranteed-maximum) scheme, except that the contractor is normally reimbursed for all his costs at a minimum. In some cases, however, cost beyond a certain amount may not be fully reimbursed. At that time the contractor may experience a loss (represented by a negative fee). The following factors are determined by the contracting parties.

1. Target cost
2. Target fee
3. Minimum fee ⎫
⎬ optional
4. Maximum fee ⎭
5. Fee adjustment formula

After performance is completed, actual costs are compared to the target cost, and the contractor's fee is determined through the fee adjustment formula. The general relationship among these factors for a typical contract is shown in Figure 10.

Cost risks to owners are similar to those under the guaranteed maximum method. Additional risks arise when costs increase beyond the minimum fee level; then, the owner assumes all costs, even though the contractor does not gain additional fees. More detailed owner cost monitoring is required due to the absence of a ceiling price. Extensive effort is needed to set reasonable cost factors prior to award and, as with all cost-reimbursable methods of pricing, postaward monitoring and cost verification are essential.

This type of pricing structure is quite common in agreements with A–E firms and other consultants. As with the guaranteed max scheme, labor hours are often substituted for dollars when determining the target costs (target labor hours). Advantages of this pricing method are:

1. The contractor has some incentive to keep costs, or labor hours down.
2. Scope of work can be less specific at time of award than for firm-fixed pricing.
3. Agreement may be reached more rapidly than with fixed-price methods.
4. Unnecessary contingencies in bid amounts should be eliminated.

A typical agreement for engineering and design services may use this method in the following manner.

For the first 600,000 labor hours, the fee will be 80% of labor hour costs; for the next 250,000 labor hours, the fee will be 50% of labor hour costs; for the next

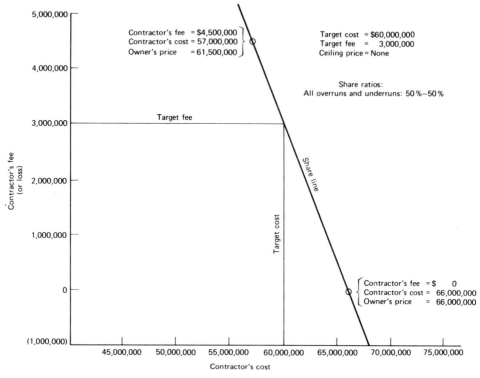

FIGURE 10. Cost-plus-incentive-fee graph.

100,000 labor hours, the fee will be 25% of labor hour costs; and for the remaining labor hours, there will be no fee; that is, any time spent beyond 950,000 labor hours will be reimbursed at actual cost—no fee will be earned for those additional hours.

Cost Plus Awarded Fee

Under this concept, the contractor is reimbursed for all costs and receives a guaranteed minimum fee plus an "awarded" fee. The awarded fee is intended as an incentive, even though its size is unknown at time of agreement. The awarded fee is subjectively determined after the work has been performed. Again, this subjectivity sometimes includes negotiation (and bitter dispute!)

The owner's cost risk is similar to that under cost-plus-fixed-fee arrangements (discussed later) because the minimum fee represents a fixed amount. But since the awarded fee is based on a subjective decision by the owner, a contractor's dissatisfaction with that amount may jeopardize future agreements. Often, the awarded fee is viewed as so uncertain as to provide little or no incentive. Establishment of the minimum fee represents the major cost-control effort during the contract formation period. And as for all cost-reimbursable methods, selection of qualified contractors and constant cost monitoring are required.

This type of pricing is used when uncertainty of work scope is too high to establish a ceiling price, or when the work is a first-time attempt and the degree of performance quality is not ensured. Seldom used for furnishing of goods or for construction services, cost-plus-awarded-fee arrangements have been used when contracting for professional services. Even under these conditions, it has limited application.

Cost Plus Fixed Fee

This alternative allows the contractor to be reimbursed for all costs and to receive a fixed fee for its services. The fixed fee is determined prior to award of the contract and is not changed unless the scope of work is revised. Since it will be reimbursed for all costs, without limit, and its fee is guaranteed, the contractor has no immediate price or cost incentives.

Because of the absence of ceiling prices, awarded or incentive fees, or cost-sharing formulas, there is less effort required on the part of the owner to negotiate or agree with a prospective contractor prior to award. However, some evaluation is required to establish the fixed fee. Constant cost monitoring and control on the part of the owner are necessary.

The fixed-fee amount should bear some relation to the degree of difficulty in performing the work, the cost of the work, and the duration of performance. And contractual agreements using this pricing method should specify the required amount of management, supervision, and control provided by the contractor. In addition, the definition of those costs that will be reimbursed and those that are assumed to be covered by the fixed fee should be thorough. Changes to the work require detailed evaluation and negotiation as to their impact on the fixed fee. This pricing method is quite common for construction contracts and for some service-related contracts as well.

Time and Materials

Under a time-and-materials (T and M) contract, the contractor is reimbursed for all materials used at their actual cost. In addition, it is paid for direct labor at a fixed rate (usually hourly). The fixed rates used for labor vary according to labor classification and include a markup for the contractor's fee. As such, direct labor hours represent the only vehicle for contractor profit and recovery of nonmaterial costs. The hourly rates should include wages, overhead, general and administrative expenses, and profit.

The time-and-materials pricing method provides no incentive for controlling material costs or labor costs. In fact, it gives the contractor an incentive to increase direct labor hours used because as these increase, its embedded fee increases proportionately.

Contracts or change orders based on the T and M pricing method are simple to obtain and can be awarded rapidly. That's why they are common for small scopes of work, such as repairs or emergencies. Greater effort is required to monitor material costs and labor hours. This method is quite common for construction work and for professional services. In general, it should be avoided for all work except that involving minor-cost, short-duration efforts that cannot be planned in advance. Always include a "not-to-exceed" limitation in hours, material, or total cost.

Cost Plus Percentage of Costs

Under this arrangement, contractors are reimbursed for all costs and given a fee that is directly proportional to some or all of the costs involved in the work. The higher the direct costs, the higher the contractor's fee. Most owners agree that this arrangement is the least favorable of all. Not only do contractors have no incentive to control costs, but they also are given a positive incentive to *increase* costs. Because of this lack of cost control, the owner's efforts during selection of the contractor and subsequent monitoring of costs and its cost-control efforts should be substantial. In many cases the fee percentage is renegotiated during the performance period, particularly if incurred costs have exceeded original estimates. This also requires considerable time and effort. Perhaps the best control recommendation regarding the cost-plus-percentage-of-costs approach is to avoid this pricing method! If it is unavoidable, set the lowest possible fee percentage and to monitor both costs and performance of the contractor very carefully. Cost-plus-percentage-of-costs arrangements are not recommended for any application. If you use them, keep them restricted to short-duration, low-cost, or emergency work performed by reliable contractors.

PROFIT INCENTIVES

Cost incentives described for several of the preceding cost-plus pricing methods represent only one form of contractor incentive. In addition to target prices, ceiling prices, share ratios, and the like, other inducements—both positive and negative—are used to promote efficiency, quality, and timeliness of contractor performance. The available incentives are grouped into the categories of cost, technical, and schedule, according to the element of performance they are intended to promote. Taken together, they are all thought of as contractor-profit incentives.

TECHNICAL INCENTIVES

Technical incentives punish or reward contractors for the quality or performance of work. They are commonly used when certain performance characteristics of a finished product are important or valuable to the owner. The concept of rewarding or penalizing a contractor according to the performance of a finished product can be applied to most contracts where design and development work is involved in addition to construction. A general principle is that it does little good to create incentives for performance that is beyond the construction contractor's control. Performance of a cooling tower, for example, is greatly dependent upon the design and manufacture of its components. To penalize or reward a contractor responsible for only the installation of the tower would make little sense.

Performance or technical incentives are loaded with the potential for dispute. When used, make sure the measurement and payment criteria are thoroughly specified and that responsibility for all aspects that influence performance rest with the contractor that will be penalized or rewarded.

SCHEDULE INCENTIVES

The term *bonus and penalty clause* usually refers to schedule incentives— sometimes called delivery incentives. These seek to reward contractors for early completion of work or penalize them for late completion. In some cases, only penalties are involved. These penalties are often referred to as *liquidated damages* and are typically stated in terms of so much money per day beyond a specified date. They have merit for projects where completion by a certain date is essential, or where significant costs are saved, in the case of early performance, or incurred, in the case of late completion, on the part of the owner.

In combining cost, schedule, and/or technical incentives into a contractual agreement, take care that the different incentives do not counteract or conflict with one another. For example, a contractor absorbing 25% of cost overruns beyond a certain target price in addition to gaining $250,000 per day for early completion would be inclined to spend tremendous amounts of money—on overtime, extra equipment, prefabricated materials, and so on—in order to secure the lucrative schedule incentive. In doing so, it would naturally forgo the 25% increase in required cost. This trade-off may be contrary to the interests of the owner. All types of incentives—cost, schedule, and technical—should be balanced when used together on a single contract. The same advice applies equally to situations where more than one contractor is working on a project. Should an earlier contractor be working under extremely high schedule incentives, it may rush its work at the expense of the interfacing contractor who follows. Many times owners adopt incentive arrangements without considering the accompanying direct and indirect effects on the contractor in question as well as others involved with the project.

PRICING CHANGE ORDERS

Regardless of the pricing techniques used for the original or "base" contract, construction contract documents invariably provide for the pricing of changes or additional work according to one of several prearranged methods. Typically, the owner allows itself the option of requiring the contractor to quote on change order work using one of three methods:

1. An agreed lump sum
2. Unit prices already contained in the contract or subsequently agreed to between the owner and contractor
3. A cost-plus-percentage fee, with the definition of allowable costs and the acceptable fee percentages already stipulated in the contract

Some owners specify the percentage fee for the third method above when designing the bidding or contract documents. Others, however, ask the bidders to insert the fee they will charge for change-order work, and, depending on the amount of extra work anticipated, this could become an important factor in evaluating bids and awarding the contract. Rather than using only one percentage markup amount, most contracts call for different markups degrading upon the element of cost involved. For example, a contract might call for the following markups to cover the contractor's overhead and profit should extra work be ordered on a cost-plus basis.

15% for direct labor
10% for material
8% for subcontracted work

Again, the owner usually retains the right to choose the pricing method, one of the three described above, by which the contractor must quote for additional or extra work. If the contract is unit priced and additional materials or labor are required, the existing unit prices may cover these additions up to a certain amount (say, 25% above the anticipated or estimated quantities). Some contracts require that any amount over or under such a range will be quoted on different unit prices.

Even with lump-sum contracts, it is wise to request bidders to quote unit prices for additions or deletions to the work should changes occur. By requesting these prices with the original contract bids, the owner receives competitive prices before it awards the contract. In their absence, it could find itself in a poor negotiating position later—when the contractor is on site and has the owner "captive." Items that lend themselves to "add-and-delete" unit pricing include tons of steel, cubic yards of earthwork and concrete, square feet of road surface, and feet of electrical cable and piping. It is not uncommon to see contractors bidding a different unit price for adding versus deleting the same items. For example, a bidder may quote an "add" price of $1.10 per pound for steel reinforcing bar and a delete price (or credit) of only $0.90 per pound for the same bar. This is an accepted practice and is based on the fact that certain fixed costs and overheads for the contractor do not decrease in direct proportion to a decrease in the amount of reinforcing steel installed.

SUMMARY

Selecting a contract type and its associated pricing strategies should be a logical process that considers the nature of the work, time of performance, contractor marketplace, completeness of design information at time of bid, and the resources of the owner to monitor and administer the contract. It goes without saying that more contract monitoring is required for a cost-plus agreement than for a lump-sum one. And regardless of the owner's wishes, it may have to settle for a different pricing arrangement if no contractors will accept the one it has chosen. Also, all contracts should have specific provisions for pricing extra work. It is preferable to include these as items to be bid at the time of original contract formation so as to take advantage of the owner's leveraged position.

One final note: It is surprising, considering all of the above, to see the number of owners who choose to ignore these considerations and select one broad pricing structure to cover all contracts on a large project—for example, "We use only lump-sum contracts." Oftentimes this is an arbitrary, illogical decision based on personal bias and inaccurate assumptions. Each contract is different, and a pricing strategy tailored to the risks associated with each particular contract is simply good business practice.

CHAPTER 4

CONTRACT PACKAGING
AND SCHEDULING

The first task the person responsible for contract formation (we shall call him or her the contract manager for expediency) must undertake at the beginning of a proposed construction project is to determine the orderly subdivision of the work into distinct contracts, or preliminary contract "packages." Basic decisions must be made concerning (1) the number of contracts to be used, (2) their respective scopes of work, and (3) schedule requirements for their bidding and award.

Contract packaging efforts, of course, depend directly on the organizational approach chosen for the project, such as design–build, force account, and multiple primes. Where a single design–build contract is to be used, this packaging and scheduling process is reduced to a simple and straightforward exercise—there is only one contract involved. With the opposite extreme, in a multiple-prime environment, it can be extremely complex and difficult. And, naturally, there are gradations in between these two examples. For purposes of presenting an exhaustive analysis of this topic, we will use the multiple-prime approach. It will present us with the most challenging and thorough illustration.

PRELIMINARY SCOPE DETERMINATION

This is the project management team's first attempt to "scope out" or define the work that each contract will represent. It should be a joint effort of personnel representing construction, engineering, and contract management as a minimum. The results of this preliminary discussion should be (1) preliminary scope documents for each anticipated contract, and (2) a milestone schedule indicating at least the following dates for each contract:

1. Contractor begins work
2. Contract award
3. Bids due
4. RFP released for bids
5. Delivery of design products (drawings and specifications)

CONTRACT SCHEDULING

Notice that the preceding milestone dates are listed in reverse order of occurrence. They illustrate that the integrated project schedule should be construction driven; that is, the contract formation and design functions should be scheduled and performed to support the needs of construction. This represents a simple "backward-looking" schedule philosophy, and the intervals between these milestones are the durations allowed for contract formation activities. This can best be shown by referring to the abbreviated critical path method (CPM) network, depicted in Figure 11. In order to obtain challenging but achievable schedules, the contract administrator must allow sufficient time for the major tasks required during the formation period.

When preliminary contract packaging and scheduling is complete, architectural and engineering groups will be able to plan their work in order to support the formation and construction processes because they will know:

1. The priority, sequencing, and grouping of its technical products
2. The delivery dates required for those products

Contract managers will know the number and scope of the required contracts, the due dates and durations for the formation tasks associated with

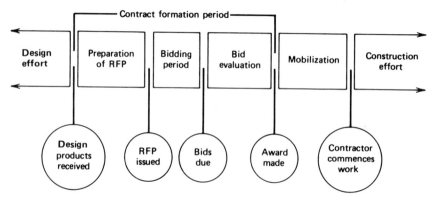

FIGURE 11. Contract formation milestones and activities.

each, and the due dates for contract award so that construction may proceed as planned. Construction personnel will know the number and order of mobilization of the project contractors, as well as what each one is expected to perform.

ITERATIVE PROCESS

This scoping and scheduling process is not a one-time effort. On the contrary, flexibility and changes will require that it be iterative and that the results constantly be updated for each contract, with the project schedule reflecting the latest dates and durations. An example of a preliminary scoping document for a particular contract—site preparation for an electric power plant—is given as Exhibit 1. It is extremely helpful for those planning the contracting process to develop such a document for each envisioned contract. When used, each contract should be defined as to scope of work, suggested pricing method, required design products, and closely related work not included.

This documentation is an extremely valuable communication tool. It tells the design group (or consultant) what support formation is required for each package, allows the contract management and construction groups to verify that all work is covered by one and only one contract—no gaps or overlaps— and allows the owner's construction organization to load its required field monitoring and support services, such as staff, field offices, craft parking lots, access, and sanitary facilities, over the project construction phase. In effect, the collection of planning documents for each contract becomes the owner's "contracting plan."

When all contracts have been preliminarily scheduled, the contract manager can determine the level of effort required at any given time in order to perform contract formation functions. This allows him or her to assign the appropriate people and to level this staffing over a period of time—to reduce peaks and valleys—if possible. A typical CPM network segment showing the durations and relationships of the major contract formation activities they will perform is shown in Figure 12.

INTEGRATED SCHEDULE

Contract formation activities represent the locus of intersection of the preceding engineering–design effort and the subsequent construction effort. This planning forces these two, sometimes disparate, schedules to join in an overall integrated project schedule. The contract packages themselves form a series of time-phased "bridges" between these two schedules. Sufficient planning of each formation step and use of reasonable durations for each activity help prevent these bridges from becoming bottlenecks.

EXHIBIT 1.   Contract Packaging Scope Document

CONTRACT PACKAGING SCOPE DOCUMENT

Contract Identification   C–11   Site preparation

General Description of Work

Preparation of construction site, including clearing and grubbing, grading, cutting and filling, installation of storm drainage and erosion prevention systems, construction of sedimentation pond, access roads and parking lot subgrades, preparation of material laydown area, and performance of initial slope seeding.

A. Scope of Work

1. Clearing and grubbing site vegetation
2. Excavation, grading, and soil compaction
3. Preparation of fill storage and disposal areas
4. Transfer of soil, rock, and other excavated material to storage and disposal areas
5. Reclamation of excavated material for use as fill
6. Transfer and disposal of waste material to land fill area
7. Hauling and placing of fill material
8. Local dewatering of contractor trenches and other excavations
9. Initial seeding of slopes where required
10. Installation of storm drainage pipe, prefabricated culverts, and catch basins (to be furnished by owner)

B. Technical Specifications Required

No. 122 Earthwork
No. 145 Site drainage system
No. 232 Landscaping

EXHIBIT 1. *(continued)*

C. Drawings Required

    1. Layout:
       Site plot plans
    2. Civil:
       Mapping and property
       Subsurface exploration
       Site preparation
       Sanitary waste system
       Dams and reservoirs
       Site improvements
       Yard structures and buildings

D. Closely Related Work Not Included

    1. Excavation of foundations for temporary or permanent buildings
    2. Concrete construction, including material supply
    3. Furnishing of drainage pipe, culverts, or catch basins
    4. General site dewatering (not including contractor excavations)
    5. Installation of subgrade piping (except for storm drainage system)
    6. Construction of roads, parking, or laydown areas above subgrade

E. Recommended Pricing Structure

    1. Lump sum (firm fixed price without escalation)
    2. Unit prices for addition and deletion of quantities of:

$$\left.\begin{array}{l} \text{rock} \\ \text{soil} \end{array}\right\} \text{excavation, fill, disposal}$$

F. Contract Formation Dates

    1. Begin construction June 1, 19XX
    2. Award contract May 1, 19XX
    3. Receive bids March 15, 19XX
    4. Release RFP February 1, 19XX
    5. Receive drawings and specifications January 10, 19XX

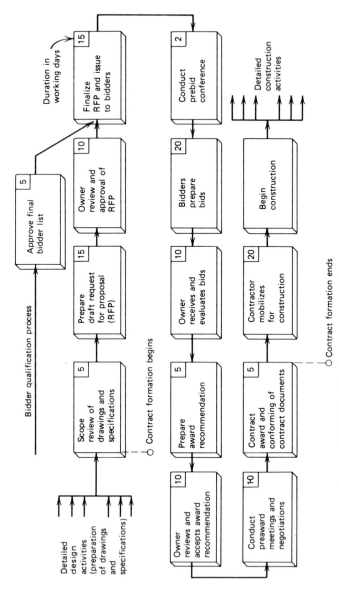

FIGURE 12. Sample contract formation CPM network.

CONTRACT IDENTIFICATION

It is customary to number each contract for ease of reference and tracking purposes. Most organizations will simply number them chronologically beginning with the first to be awarded and continuing to the final contract (C–1, C–2, C–3, and so on). Often, prefixes will be attached to the numbers to indicate engineering discipline or construction trades associated with each contract. For example, two electrical contracts are called E–1 and E–2; mechanical contracts, M–1, M–2, M–3, and so on. Long, convoluted multicharacter identifications are strongly discouraged, for each will be used thousands of times throughout the contracting process.

REPRESENTATIVE CASES

1. A project for constructing an 80-story office tower and adjacent parking garage was in process. A multiple-prime contracting approach was being used, and the issuance of the RFP for furnishing and erecting structural steel was being delayed because the steel design drawings were incomplete. After several weeks, the lead structural engineer, tired of being made a scapegoat, exclaimed that she was being held up by the indecision on the part of the owner's building manager regarding the choice of structural steel versus precast concrete for the elevated parking decks. In actuality, the parking garage was never intended as part of the scope of work for the structural steel contractor—even if it were to be made of steel! The garage was singled out as a separate turnkey contract much earlier, but this decision was unknown to the lead structural engineer. A contract scoping document and her participation in the planning of contract packaging would have prevented this.

2. The chief of contract management reviewed the integrated project schedule and decided it was ambitious but achievable. Durations given for the formation of each contract were reasonable. He failed, however, to consider the combined effect of all contracts on the manpower in his department at any given time. Murphy's Law prevailed, and he later found his three contract administrators scheduled to conduct six prebid conferences, issue four RFPs, and evaluate 12 contract bids during the first week of April! During the remaining three weeks of that month, they had no assigned work, so they spent their time grumbling about their leader and updating their résumés.

3. Because he knew that a specifications writer was to be hired in several months, the lead mechanical engineer decided to concentrate his staff's efforts on mechanical design and drawing production during the intervening period and to defer the writing of contract specifications until the new man came aboard. The drawings for all five mechanical contracts were completed ahead of schedule, but two RFPs were severely delayed due to lack of technical specifications.

4. The electrical drafting squad leader made a meticulous takeoff from the master integrated project schedule at the beginning of the project. She created a separate, detailed drawing production schedule that, if followed, would ensure that his drawings would be ready for each contract RFP when needed. During the next several months, work on the chemical plant's generating unit was delayed due to permit problems, the cooling tower was accelerated by 10 weeks to escape a pending environmental deadline, and purchasing lead time reports showed transformers requiring six months for delivery instead of the original three. All this was reflected on the updated project schedule, but since the drafting squad leader had been operating from her personal "pocket schedule," her production priority was totally out of synchronization with project needs.

PART 2

CONTRACT FORMATION

The previous chapters dealt with assigning responsibilities and selecting pricing and scheduling controls for an entire project; this part addresses the steps to secure fair and equitable contracts that (1) are well defined and priced, (2) protect the interests of each contracting party, (3) contain enforceable compliance controls, and (4) can be properly administered.

Each of the following chapters describes steps taken in a sequential manner, for each contract, to achieve these objectives. These steps consist of:

Developing Contract Documents. The first step is to choose or create project- or company-wide documents that will form the basis of control for each contract-specific application.

Bidder Qualification and Selection. This formal, structured process is aimed at reducing bid evaluation effort, soliciting competitive bids from the best possible offerings, and preventing performance problems once construction has started by allowing only responsive and qualified contractors to participate in the bidding process.

Issuing Requests for Proposals. The general reference or standard documents are tailored into specific bid packages. Steps taken to prepare each RFP will be reviewed and operational considerations offered to help achieve well-priced, clearly scoped, and tightly bid contracts.

Managing the Bid Cycle. The selection of the best bid offering begins long before bids are received. In this chapter we describe ways to channel bids into acceptable offerings by controling bid information prior to receipt with addenda, prebid conferences, bid extensions, and contractor site visits.

Bid Receipt and Evaluation. Methods must be followed to control the evaluation process, maintain bid security, and ensure that the cost, technical content, and commercial considerations of each bid are thoroughly analyzed so that the best possible offering is accepted.

Contract Award. Once a successful bid has been identified, specific steps are needed to conform the documents into enforceable and easily administered criteria. The use and abuse of letters of intent, as well as the incorporation of changes and understandings made during preaward meetings and negotiations described.

These steps will take us from the point where the need for contracting for outside services is recognized to the achievement of a signed contract, with a qualified contractor, at acceptable levels of commercial cost and risk.

Acts or omissions during this formation phase greatly affect the success of contract administration and contractor performance to follow. Formation controls are, therefore, important not only because they benefit the selection and award process but also because they set the stage, cast the players, and provide the script by which the actual contract work will be performed.

CHAPTER 5

DEVELOPING CONTRACT DOCUMENTS

Two distinct sets of documents are used by owners to communicate their needs and to structure the activities of both parties during the contract formation and administration phases: bidding documents and contract documents. For most complex projects, these are often voluminous in size, encompassing in scope, and very detailed in content. Owners have three general sources of contract and contract-related—that is, bidding—documents:

1. *What Worked Last Time.* The owner chooses the most appropriate and best documents from its files of previously used documents.
2. *Standard Documents Available in the Industry.* These include bidding and contract documents published by professional associations or certain industry groups involved with construction. Examples are those provided by the American Institute of Architects (AIA), the National Society of Professional Engineers (NSPE), and the Associated General Contractors of America (AGC), to name a few. Standard contract documents are also published by various associations for specific application to international projects. An example is the "Conditions of Contract for Works of Civil Engineering Construction" published by the Federation Internationale des Ingenieurs-Conseils (FIDIC).
3. *Reference Documents Maintained by the Owner or Its Representatives.* These include documents prepared and maintained for a range of applications. They usually contain standard clauses and terms and provide for the modification or insertion of project, owner, contract, or jurisdictionally specific terms and conditions. They are often maintained on electronic memory for easy modification.

Of these three choices, most owners choose from the final two. They realize that unless "what was used last time" is recent, was effective, and does not differ substantially from present needs, much revision is normally required to modify specific contract documents used in the past to a new application.

INDUSTRY STANDARD VERSUS OWNER REFERENCE DOCUMENTS

Although owners of major projects have used industry standard documents with varying degrees of success, these are much more commonly found in conjunction with smaller commerical or architectural construction. To determine whether industry standard or owner reference documents should be used as the starting point for developing specific documents tailored to needs of each proposed contract, owners should consider the benefits and disadvantages of each. Let's review some here.

Industry Standard Documents

These documents have several advantages:

1. They have generally been carefully designed and reflect the collective experience of many practitioners in many project applications. They offer basic controls over common contractual risks. They are updated periodically to incorporate enhancements gained through their use and to reflect changing industry practices and legal interpretations.
2. They can be obtained and used easily. The effort of creating contract documents is greatly reduced, and the time required to achieve owner-specific versions is sometimes minimal.
3. Some proponents of industry standard documents claim that they have more weight in court than those created by the owner (or contractor) without the endorsement of a sponsoring agency or association.
4. They are probably more familiar to contractors. As opposed to specifically designed documents, standard documents have been seen and used by most major contractors, who are more comfortable with standard documents during the bidding stage.
5. They are inexpensive to obtain and to use. Most are available without charge.

These standard documents do involve some disadvantages, however:

1. They generally favor the drafter (sponsoring organization). An examination of documents prepared by engineering or architectural associations compared with those prepared by contractor groups points this

out very directly. Protection is often afforded the party represented by the sponsoring association at the expense of the others.

2. They reflect assumptions that may not be merited for specific applications. Most industry documents are written to have broad application, in many jurisdictions and under widely varying circumstances. Because of this, they do not contain specific language reflecting owner needs as they may vary from one application to the other. It is difficult to isolate and revise these assumptions to reflect specific owner management and control objectives without an exhaustive analysis. Often the assumptions are implied or embedded in nonrelated clauses.

3. They reflect "typical" or assumed roles and responsibilities for major contracting parties, such as owner, architect, engineer, contractor, subcontractor, and construction manager. But these may be very different for a specific project. It is not unfair to say, in addition, that the authority is slanted toward the group represented by the drafting organization, and questions of liability or responsibility are slanted away from that group as well.

Owner Reference Documents

Owner reference documents have advantages over industry documents:

1. They can be more precise and specific regarding owner, project, and jurisdictional settings.

2. They can implement specific control objectives that may not have been envisioned by the drafters of generic documents.

3. They reflect known rather than assumed roles and responsibilities.

4. They can contain owner protection not afforded in different jurisdictions. For example, should a hold harmless clause favoring the project owner be illegal in a majority of the states, industry documents will most likely omit this protection. If the owner and project in question are not bound by such legal restrictions, reference documents may incorporate the protection offered by such a clause.

5. They can match other owner-prepared documents more closely than can generic versions. These include any other bidding and contract-related forms, instruments, and reports (change orders, technical specifications, backcharges, and so on).

However, they have disadvantages as well:

1. They are more expensive to obtain. Much time and effort is required to create and maintain owner reference documents.

2. They may contain clauses whose enforceability is untested.

3. They may be unfamiliar to contractors. Bidders and contractors may be uncomfortable with or suspicious of the terms and conditions proposed by the owner. Higher prices could result.
4. They may be poorly written or incomplete. The quality of reference documents is a function of the knowledge and ability of those who prepare them.

The choice of one document source over another must be made with full awareness of the benefits, costs, and risks pertaining to each, and in full view of the specific circumstances surrounding project contracting applications. For most owners of large industrial projects, reference documents are created during the initial planning stages and are revised to reflect the specifics of each anticipated contract. This chapter will assume this to be the case when discussing the documents themselves and the operational considerations involving their use. Again, this gives us the most exhaustive illustration.

Standard documents are common, but their effectiveness depends upon the owner's understanding of their contents and a careful tailoring of their general terms to match specific needs and control objectives. In any case, both types of documents should strive for consistency and uniformity and each represents a good starting point for contract-specific modification— which is always required, no matter the source of the original documents to be used.

CONTRACT AND CONTRACT-RELATED DOCUMENTS

Bid packages generally contain most or all of the following types of documents (examples of each are given as exhibits in Chapter 7):

1. *Invitation to Bid.* This is usually in letter form and sent to bidders inviting their proposals and explaining the rules of the bidding process.
2. *Proposal.* This is designed by the owner and completed by the bidder with such information as prices, schedules, and submittals. It references the remaining four sections and the drawings as the basis for the ultimate contract.
3. *Agreement.* This section is executed by the contracting parties once a bid is accepted.
4. *General Conditions.* General requirements that apply to all construction contracts for the particular project are given here. The basic relationships between the contracting parties, general project rules, and commercial terms are contained in this section.
5. *Special (Supplementary) Conditions.* Particular requirements of work to be covered under the contract are given here. These include scope of work, material and services furnished by the owner or others, special

site conditions affecting access, laydown areas, storage and utilities, and limits of responsibility of the contractor.

6. *Technical Specifications.* Prepared by the responsible design organization, these give detailed technical requirements of the work to be accomplished. Codes and standards, performance and acceptance criteria, and materials and methods are prescribed. The technical specifications are a group of documents (or divisions), each pertaining to a particular installation or service involved in the work.

7. *Drawings.* Prepared by the responsible design organization for the owner, these graphically depict location, size, shape, and details of construction or composition of the work.

The preceding items comprise the bid package or request for proposal (RFP) package sent to bidders. Once a successful bidder is selected for award, items 1 and 2 are withdrawn, and a conformed contract containing items 3–7 is prepared and executed. These items are called the contract documents (see Figure 13).

CONTRACT DOCUMENT DESIGN PHILOSOPHY

For smaller commercial and architectual construction projects, many owners and design professionals prefer to adopt standard industry-endorsed bidding and contract documents, such as AIA documents, and modify these to the particular needs of the owner, contractor, and project conditions. In a parallel effort, the Construction Specifications Institute (CSI) has developed detailed technical specifications formats and guidelines. When properly and appropriately used, these documents and guidelines greatly reduce the docu-

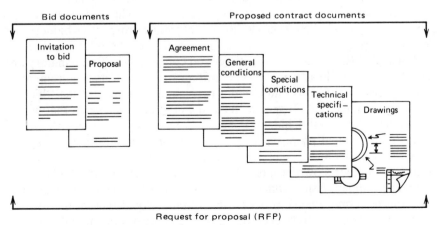

FIGURE 13. Typical bidding and contract documents.

ment creation process, contribute to uniformity among contracts and projects, and result in high-quality documents.

There is much debate, however, among those who deal with large engineered projects—power plants, refineries, transportation systems, heavy industrial facilities, and so on—as to their application in these environments. A general consensus, if one exists, is that without considerable modification the generic, industry standard formats are not especially suited to these types of projects. Those responsible for preparing contract documents (including specifications) in situations where available standard documents are deemed unsuitable are required to design and create their own. Often this is done on a project-by-project basis. However, repeat builders and engineering or project management consultants have recently begun to produce "reference" contract documents that can be modified quite easily to suit the requirements of a particular owner, legal jurisdiction, project type, and organizational–contractual approach.

Regardless of the origin of the bidding and contract documents, there are certain guidelines, or caveats, that should be followed to streamline the processes they affect: contract formation and administration.

CONTROL OF THE DRAFTING PROCESS

One of the most foolish actions an owner can take is to request proposals from bidders without specifying the format and content it expects to receive. In effect, it is saying, "Bidders, send us whatever you choose." It is easy to imagine the menagerie of different proposals that will surely result. The bid evaluation process is difficult at best, and chaotic as a rule. Comparing widely dissimilar proposals would resemble comparisons among apples, oranges, bananas, and bricks.

Most astute owners (or their representatives), therefore, channel the bidders into conforming to particular proposal formats. This is done by using a properly designed bid package (RFP) and informing prospective bidders that deviations from them may be cause for rejection of the offending bids. A well-designed RFP is simple and direct and completely self-explanatory and reflects a reasonable attempt to avoid misconstruction of its requirements.

RFPs designed to this end contain two major sections: (1) the "bidding only" documents; and (2) specimens of the contract documents that will be used once a successful bidder is selected for award. Bidders should be furnished a detailed proposal form, soliciting specific information, and required to use this form to submit the details of their bids. Use of "bidder's paper"—a bidder's own proposal format, standard conditions, and so on—should be avoided. Aside from complicating and extending the bid evaluation and contract conformance process, this offense would require the owner's staff to review completely each detail of each bidder's documents for each contract (a lot of fine print to squint at). It cannot be overemphasized that the owner must maintain authorship and control over the bidding and contract documents. From the postaward perspective, contract administration is difficult even with several contracts containing identical commercial and legal terms. Were each of these to be significantly different, the contract administrator would soon become convinced that he or she is working on a project to build the Tower of Babel.

Bidders have a knack for inserting their own clauses, deleting existing ones, and qualifying their bids through other means. Again, to have to check for interlineations, erasures, typeovers, and marginal notes is an unbearable and risky task. Each of these should be prohibited and bidders required to state exceptions to the provided documents on a separate sheet of paper, bearing their letterhead, with suitable references to the specific items as addressed in the owner-furnished documents.

Some owners will award contracts with a letter stating "on the basis of your proposal dated" or, rather than transferring bidder quotations from the proposal form to the agreement form before the contract is signed, will instead incorporate the bidder's proposal into the contract by reference. This is dangerous and should be avoided. It takes little time and effort to transfer the quoted prices and other variable information—names, dates, and so on—to the owner-designed agreement form. The bidder's proposal can be filed away. To facilitate this process, the proposal and agreement forms should be parallel in design.

OPERATIONAL CONSIDERATIONS

In addition to the general guidelines given above, there are several specific suggestions that will aid the formation process. For example, where certain documents are required for all contracts of a project and are not subject to modification, consider preprinting several hundred copies to save reproduction costs and time. These might include the following:

General conditions
Insurance requirements
Project agreement
Job rules
Safety program requirements
Special instructions (e.g., drawing submittal requirements, or quality assurance program).

DOCUMENT RELATIONSHIPS

Document hierarchy (which documents should dominate if conflicting requirements are discovered, and so on) should be fully described in the documents themselves. A general rule is that written instructions, requirements, and so on, should occur only once and in the proper documents. The drawings are issued with the bid packages (RFPs) to give the bidder (and later as "contract drawings" to give the contractor) information that can be transmitted more effectively in graphic form than in written language. For this reason, instructions that are able to be written should be placed in the technical specifications rather than in the drawings. Anyone who has been through this process knows how much more difficult, time-consuming, and expensive it is to revise and reissue drawings as opposed to modifying printed contract documents when changes occur. People preparing the drawings and those preparing the written documents need to communicate to avoid redundancies and conflicts between the two.

This same principle applies to the technical specifications and their relation to the other written documents (general conditions, special conditions, and so on). A common practice is for engineering departments to author the specifications and drawings while contract managers prepare the remainder of the documents (commercial "boilerplate"). Each should be aware of the topics it is to address. Often, the technical specifications contain commercial and legal terms or instructions embedded in the technical requirements. This creates three apparent problems:

1. There is a strong possibility that they are in conflict or redundant with the commercial sections.
2. If technical specifications remain only technical, they can be readily used for other projects or other contracts without reissue or modification. If they contain project-related terms and conditions, this efficiency is forgone.
3. Generally speaking, people trained to create technical specifications are not qualified to decide on or address commercial and legal questions.

The same is true for the drawings. Since the same drawing may be used for several separate RFPs and contracts (with each contractor performing only certain portions of the work), the scoping of the work among different contractors should be done through the special conditions section. Notes on drawings such as "by owner," "by others," or "by C–15 contractor" are discouraged. These scopes will change as the project progresses, and again, it is much easier to reallocate scope through written documents than to redraw, check, approve, reissue, reproduce, and distribute drawings with scores of copies each to several affected contractors and internally among the owner's and designer's organizations. Why ask for this nuisance?

USE OF PURCHASE ORDERS

Many owners issue purchase orders to cover virtually all items they buy. Accordingly, they issue a purchase order to procure contractors' services, or contracts. This is acceptable, providing the purchase order does not contain standard or preprinted terms and conditions (most of them do). The problem lies with these terms and conditions. They are intended to cover commercial *purchases*—that is, office supplies, raw materials, fuel, and so on—and do not suit the special requirements of construction contracting.

A basic tenet is that purchases involving job-site labor (with or without the furnishing of material or equipment) should be accomplished through a construction contract and not a purchase order. If owners insist on issuing purchase orders (POs), the PO should merely reference the applicable contract—one that has been prepared and formed in accordance with the recommendations contained in this book—and should specifically state that all printed conditions on its face (or reverse) are null and void.

The use of POs is an acceptable practice, but when used in conjunction with construction contracts, they should be reduced in significance to that of a tracking and transmittal vehicle, not as a replacement or supplement to the contract documents. For the same reasons, change orders to contracts should stand alone, and PO amendments or modifications should not replace or supplement these either.

DOCUMENT MODIFICATIONS

Changes to contract or bidding documents take two forms: (1) changes to the bidding documents (RFPs) during the bid cycle (called *addenda*); and (2) changes to the contract documents once the award is made (called *change orders*). An intermediate state exists when an accepted proposal is being incorporated into the final contract documents prior to their signing (or execution). No formal instruments are required for this process. Addenda are issued to bidders during the bid cycle (while RFPs are "on the street"). They should be numbered, dated, and tightly controlled. This is also true for change orders. Both will be discussed in detail in later sections.

The point of introducing these two modification instruments here is to distinguish them from each other and to emphasize that in order to maintain control over both RFPs and contracts, strict adherence to these principles should be enforced:

1. RFPs should not be amended unless by addenda.
2. Contracts should not be amended unless by change order.

Some owners accept the use of a plethora of change instruments. These include such things as contract modifications, amendments, extra work orders, field orders, contract guidance and intent statements, and bulletins. Problems in controlling and determining the status of RFPs and contracts and processing contract transactions—to name a few—increase geometrically with the introduction of each additional instrument of change. In order to maintain a disciplined and controlled approach to contract formation and administration, addenda and change orders, when properly used, are all that are needed to implement change.

ANCILLARY CONTRACTS

There are occasions when the format or content of documents designed for major construction contracts are not well suited. Three major examples are:

1. *Short-Form Contracts.* These are designed for rapid deployment from a field site location. They cover short-duration, low-cost activities. They should be designed and issued under the same guidelines used for the formation of major construction contracts. But due to the nature of the work (snow removal, temporary access roads, and so on) and for expediency, they are streamlined, condensed versions. Short-form contracting is discussed in Chapter 16.
2. *Equipment Rental Contracts.* When fully operated and maintained construction equipment is rented, special terms and conditions apply. Bidder selection, proposal submission, and contract preparation are essentially the same as for short-form contracts.
3. *Technical or Consulting Services Contracts.* Specific terms are applicable for such services as architectural or engineering services, inspection and testing services, and surveying. Topics important to construction contracts are not required, whereas specific terms relating to patents, professional licenses, and liability, among others, are appropriate.

If a significant amount of any of the instances above is envisioned, it may be advisable to prepare standard or "reference" documents for them as well.

BONDS

There are two major categories of bonds in use today for commercial transactions and relationships. These are (1) fidelity bonds, which offer protection or indemnification against dishonesty on the part of the principal (bonded person or organization); and (2) surety bonds, which protect against loss due to the principal's failure to perform under a contract or to otherwise not fulfill its contractual obligations. It is the latter category that is widely used in conjunction with construction projects.

Surety bonds differ from insurance protection in that they bind the surety (bonding company) to make good any loss of the obligee (owner, in most cases) caused by the principal's, or contractor's, failure to perform. There are four types of surety bonds used frequently to protect project owners:

1. *Bid Bonds*. The surety guarantees that if the named contractor's bid is accepted, the contractor will indeed sign the contract and secure any performance or payment bonds required. A bid bond ensures not contract performance but only contract acceptance.
2. *Performance Bonds*. These protect the owner, up to a specified bond penalty amount, against the contractor's failure to perform its contractual obligations. They are typically secured at time of contract award.
3. *Payment Bonds*. These protect the owner against loss should the contractor fail to pay its subcontractors or suppliers. They are designed to protect against mechanics' or materialmen's liens against the owner's property.
4. *Maintenance Bonds*. These bonds bind the surety (and contractor) to correct defects in the construction work that appear beyond contract closeout up to a specified time limit. An example would be a 10-year bond against leaks in a roofing system.

Most construction contracts awarded by the U.S. government are required to be covered by performance and payment bonds (the Miller Act). Most states have adopted "little Miller Acts" that require these bonds for state projects. Where these bonds are not required by law, project owners are free to choose whether or not to require bonds.

INSURANCE

No owner would knowingly tolerate the presence of an uninsured contractor operating on its project. The risks inherent in construction activity are too great for the contractor and the owner to omit adequate insurance coverage from the contract requirements. Most of us carry all types of insurance and are aware of the complexities of policies regarding medical, life, homeowners, and automobile insurance. Contractors carry insurance on their business and operations, and understanding all of the intracacies of construction insurance is a difficult task beyond the scope of this chapter. It is beneficial, however, to understand the basic insurance coverages applying to the construction industry and common ways that project owners protect themselves through insurance.

Virtually all construction contracts require contractors to maintain minimum levels of coverage in certain areas. There are two principal reasons for this:

1. Should uninsured liabilities arising from an accident or occurrence affect the financial position of a contractor working on the project, the project could be in jeopardy or delayed.
2. The owner could be named as defendant or held responsible for damage arising from contractor operations.

As evidence of coverage, owners require contractors to submit "certificates of insurance" signed by a representative of the insuror and certifying that minimum insurance coverages, as required by the contract documents, are in effect. The insurance industry has adopted a standard form for such a certificate—an "accord" form.

Unless the nature of work a contractor will perform has particular hazards requiring special coverage (such as aircraft operations, or construction on or near bodies of water), the following four types of coverage are generally required:

1. Comprehensive general liability
   a. Operations–premises liability
   b. Independent contractor's protective liability or owner's protective liability
   c. Completed operations and products liability
   d. Contractual liability
2. Comprehensive automobile liability
3. Worker's compensation and employees' liability
4. Umbrella excess liability

LANGUAGE CONSIDERATIONS

Nothing is more tedious and frustrating than wading through a morass of obfuscating contract language. Construction contracts and RFPs are intended to transmit precise information to people who must act on it—and that action results in concrete (literally!) products that are extremely difficult to modify. Contract and specification language should be clear, concise, and direct.

The following guidelines are given in lieu of a detailed treatise on contract writing:

1. Avoid "legalese." Unless they are absolutely needed for clarity, omit words such as *hereinafter, said* (as in "said party shall . . ."), *aforementioned,* and *herewith.* Avoid pompous phraseology, such as "the party of the first part, at which time and upon said notification, as pertains to the rights, duties, and obligations of one engaged in such an endeavor."
2. The use of titles or terms interchangeably should be avoided. Although one frequently sees *owner, company, buyer,* and *company's representative,* or *contractor, vendor, seller,* and *supplier* used to refer to the owner and contractor, respectively, this should be avoided. Throughout the documents, one and only one term should be used to identify a distinct party or item. Similarly, the terms *drawings, plans, contractor's drawings, shop drawings,* and so on, should not be used to

identify the same thing. Each term has a defined, contractual meaning and should be used accordingly.

3. Avoid the common temptation to repeat requirements. Say it once, say it where it should be said, and forget it. When the same requirement is stated in several places among the documents, besides being a nuisance to the reader, changing that requirement—finding it, making sure you found each reference to it, and changing each one—invites unnecessary risk and effort.

4. Use each document for its intended purpose. Do not put technical requirements in the general conditions or commercial terms in the technical specifications or on the drawings.

5. Review and update standard or reference clauses and documents periodically to reflect changing needs, legal interpretations, governmental requirements, industry practices, and organizational preferences. Do not use 20-year-old documents even though "they seemed to work well last time."

6. Anticipate problems, misinterpretations, and scope changes and provide for these in the contract documents.

7. Get it in the contract! Contractors cannot be expected to read minds or anticipate and provide for particular owner requirements. If you want something, state it in the RFP and contract documents.

8. Consider use of the word *shall* to indicate required action or results of the contractor. Use the word *will* when describing activity by the owner or others. This helps clarify scope and responsibility assignments.

Additional suggestions are:

1. Consider standardization of technical specifications for a project, or, ideally, for all projects.

2. Do not ask for unusually complex or unnecessary bidder submittals (schedules, résumés, equipment lists, and so on) where they are not required for bid evaluation or contract administration.

3. Do not attach "reference" or "supplementary" specifications without defining which portions of such are applicable to the contractor's scope of work.

4. Number each paragraph of the commercial and technical documents for ease of reference during the bid cycle and contraction administration. Number pages as well.

5. Avoid excessive use of drawing notes in lieu of requirements in the specifications. Anyone who has been involved in revisions to bidding and contract documents will attest that drawings are much more difficult to revise and reissue than is printed material. Also, confusion and error may result when drawings containing contract-specific notes

(such as "by owner," "by others," or "by C–15 contractor") are issued to contractors or parties other than those for whom the notes were intended.

REPRESENTATIVE CASES

1. A time-and-materials contract was bid for surveying services for a mass transit project. Bidders were requested to furnish hourly rates for standard crews including equipment, transportation, labor, material, and all overhead, profit, and other fees. For the eight-person crew, the bids were:

Bidder A: $240 per hour
Bidder B: $210 per hour
Bidder C: $232 per hour
Bidder D: $245 per hour

Since bidder B was obviously low and all were well qualified, a contract was prepared referencing bidder B's proposal. At the first progress payment meeting, contractor B presented an invoice as follows: 160 hours at $294/hr. = $47,040. The owner's contract administrator took immediate exception to the hourly charge, noting the contractor's proposal stated $210 per hour, not $292 per hour. The contractor then took a copy of the contract and showed in three places where its proposal (in fine print) clearly stated "a 40% markup is to be charged to quoted hourly rates to cover general and administrative costs." Further, its proposal stated that "fees will be determined in accordance with our standard procedure," which, as described in the fine print, included the 40% markup in making the hourly fee $210 × 1.4, or $294 per hour.

2. For a highway construction project separate contracts were to be let for roadway lighting. Since the lighting contractor was originally not to install concrete standard bases (pole foundations), all drawings showing these were marked "not by electrical contractor" with a hand-drawn circle around each base. At the prebid conference, most bidders expressed a cost savings to the owner if bases were included in the standard installation. An addendum was issued to include the base installation as part of the scope of work. Electrical drafters, eager to make the change, spent five days removing the circles and notes from 188 drawings. When they had finished, the project owner reversed itself and ordered the bases removed from the scope of work again.

3. When preparing piping diagrams and drawings, the mechanical engineer noted on each affected drawing that pipe stands were "by others"— thinking that the mechanical contractor would not be responsible for stands since this work was for the civil contractor. When the civil contractor was later given the same drawings, it claimed the work was not included in its bid since the stands were clearly marked as "by others" on contract drawings.

4. Concerned that a test fill operation for the compaction of soil at a critical foundation be highly controlled, the specification writer laced the technical requirements with phrases such as:

"The work will be under the constant supervision of the owner's representatives"
"The owner will supervise and direct the test fill program to ensure satisfactory results"

At the same time, the contract manager was preparing general and special conditions disclaiming any owner responsibility for supervision in order to maintain the independent contractor status of the successful bidder. While the RFP was on the street, a large personal injury lawsuit was concluded concerning the owner's previous project. The court denied liability on the owner's part from the injured worker's claim by upholding the independent contractor relationship and citing the absence of the right by the owner to supervise the contractor's work. In his remarks, the trial judge allowed that had the owner directly supervised the contractor's work on the previous project, he could be held responsible for the worker's safety and therefore liable for his injury.

5. At a contract closeout party (after he had received and cashed his final check), the contractor's job superintendent allowed that he brought the lump-sum job in for his boss at $200,000 below the estimate. Later he divulged his technique in a spirit of greater conviviality: standard specs calling for hot-dipped galvanized piling were included in the RFP and the additional cost was included in his company's bid. After receiving the award and before ordering a piling, he found the galvanizing was not required—it was simply a standard specification clause the owner forgot to delete. He then left the party in his new luxury automobile, claiming that the owner left at least $150,000 on the bid table.

CHAPTER 6

BIDDER QUALIFICATION AND SELECTION

There are two principal methods for soliciting bids: *invitation* and *advertisement*. Each constitutes notification by the owner that a future contract is to be awarded. Many publicly funded projects require that all potential bidders be notified and allowed to submit bids. This is commonly done under precise government or agency regulations, with an announcement of the upcoming bidding process publicized in newspapers, trade journals, or other media. As such, the initial contact between the owner and potential bidders is gained through "advertisement." In the public sector, however, bidders are usually notified on a selective basis; that is, only those contractors from whom the owner wishes a bid are notified.

Given the choice between this invitation process and the advertisement method, most owners would choose the former—they would prefer to select potential bidders rather than to leave the bidding process open to the public at large. This chapter contrasts control methods applicable to bidding when performed under advertised or invited purposes, discusses the bidder qualification effort encountered with the invitation process, and presents recommendations for owners when requesting bids by invitation.

INVITATION VERSUS ADVERTISEMENT

Because advertising for bids is often required for public projects, the government or agency for which construction is performed (or under whose jurisdiction bidding takes place) usually prescribes a specific bidding and bid evaluation procedure. All bids are normally welcomed, and the opening of bids is commonly done in public, with the content and amount of each bid becoming a matter of public record.

Because the process is dictated by regulation or statute, the persons responsible for soliciting bids and for eventually evaluating them are generally restricted as to what may or may not be done during the bidding period. For owners in this position, the process is simply one of following the rules established by others. And because the project usually involves the expenditure of public funds for the public good, great emphasis is placed on the price of submitted bids, often at the expense of other characteristics of the bids and the qualifications of the bidders themselves. As such, price inevitably becomes the governing criterion in selecting a successful bidder. Because bids will be received from virtually any source, the ensuing evaluation is often time-consuming and difficult. It is also difficult to control the quality of the proposals themselves as well as the quality of those who have submitted them.

When bidding is done by invitation, however, only those bidders prechosen by the owner are allowed to submit proposals. So the owner must go through a qualification process—a detailed review of the capabilities and quality of potential bidders before choosing who will be asked to submit bids. And because bidding is done privately, confidentiality becomes an objective of all involved. This confidentiality concerns the identity of those who are allowed to bid as well as the results of the bid evaluation itself. Because most private owners who use the invitation method are totally free as to whom they may award a contract, the low bidder (in terms of price) need not always be given the award or even considered for the award.

Other factors of interest to the owner enter into the award decision; factors such as financial stability, performance on previous contracts, adequacy of management, supervisory and craft personnel, and ability to control performance are often major considerations. Because the owner is in charge of bidding conditions and free to award to whomever it chooses, the bidding and evaluation processes are generally much more easy to control than if conducted under advertised conditions.

THE QUALIFICATION PROCESS

This is the bidder qualification or prequalification effort leading to a contract bidder list. There are several reasons for qualifying potential bidders before allowing them to bid:

1. *To Identify Stable Contractors.* Few things would be worse than allowing an unstable contractor to perform contract work. Whether it is financially unstable, has a history of poor performance, or has demonstrated a penchant for disputes and claims, it is much better to discover this beforehand and prevent this contractor from bidding than to deal

with it as a project contractor. The major purpose of prebid qualification is to eliminate these contractors from consideration early.

2. *To Prevent Contract Performance Problems.* If a contractor is unsuitable for the work in question (even if it is stable and has a good overall performance reputation), it should not be allowed to bid. It may not have the resources to perform the particular job (people, equipment, funding) or have experience in work this contract requires.

3. *To Ease the Evaluation Process.* Bidder qualification eliminates the need to evaluate bids from contractors who would never be considered for award, regardless of price. Those evaluating bids may then concentrate on genuine offerings made by qualified contractors.

4. *To Obtain the Positive Benefits of Contractor Competition.* The previous reasons may indicate that the process always reduces the number of potential bidders. On the contrary, bidder qualification should be performed with the opposite goal in mind—to expand the list to the number of bidders and get competitive bids. If owners exhibit a major fault here, this may be it—the failure to seek the largest number of qualified bidders possible.

QUALIFICATION PROCEDURE

A sound qualification process begins by indentifying a large number of potential bidders. This list may draw upon published registers of available contractors, of which there are many available. A preliminary bidder list is made from the larger source list, and the included contractors are reviewed as to their quality standards, past performance, financial standing, and other factors deemed important to the owner.

The owner or its representatives then contact prospective bidders to determine their interest and intention to bid. References are contacted to determine if the candidates are financially and technically qualified. These references usually include banks, sureties (bonding companies), and previous owners or general contractors for whom the candidate has performed. To facilitate this process, many organizations maintain continuously updated bidder files. They may also prepare and issue standard questionnaires to contractors from time to time to solicit specific information that could bear on their qualification (such as financial statements, union affiliations, construction equipment inventory, geographical territory, and size of jobs they can undertake). These files are easily automated—with contractors indexed and sorted according to numerous factors in a data base. Standard, preprinted reference forms also help document telephone responses from references. Often references hesitate to disclose confidential information, but a call to the bidder candidate (and his or her subsequent calls) usually jars this information from a recalcitrant source.

A common error is to allow bidder questionnaires, reference documentation, or even bidder lists themselves to languish for months or years without being updated or reviewed for changes. For any qualification to be effective, it should be based on recent information. Procedures should call for periodic revision of bidder qualification files and bidder lists.

Another problem concerns the rigidity with which company bidder lists are often used. Some companies attempt to establish a company-wide bidder list—covering many different scopes of work, geographical locations, price limitations, and the like—but find that their overall list is so restrictive that it is virtually impossible to find competitive bidders in any number for a specific job. And it is entirely possible that any one bidder may be qualified to perform certain work yet totally unqualified to perform others. For these reasons, general, all-encompassing bidder lists should be avoided. Instead, company-wide source lists may be used, with potential contractors taken from these lists and further qualified according to the specific requirements of the job in question.

INTERNAL SOURCES OF BIDDER INFORMATION

One valuable source of information is often overlooked by owners and their project or construction management consultants. This is the performance by the contractor on previous or concurrent projects for them. To avoid this, periodic performance reports should be issued for each contractor while it is under contract and a simple performance appraisal (see Exhibit 21) should be completed and sent to the bidder files when each contract is closed out. It is surprising that in our haste to find out about a contractor, we often neglect internal sources of information.

THE TIMING OF QUALIFICATION

The degree of qualification required before allowing a contractor to bid is a subjective determination. The same is true for the proper number of bidders needed to ensure a representative sample of prices and the lowest available cost. Arguments can be made for a perfunctory (or no) qualification prior to bidding since all but one of the bidders will be rejected—that only the apparent successful bidder need be investigated. Problems with this approach are:

1. Time may not permit adequate qualification once bids are received.
2. If bidders are obviously unqualified, why waste time and effort receiving and evaluating their bids?
3. Compromised standards of quality may be more readily adapted in view of low bids (price may unduly influence judgment).

4. The owner may discover that *no* bidders are qualified.

5. It could be embarrassing or require unnecessary documentation, approvals, and justification to award the contract to someone other than the lowest bidders simply because they are not qulified.

6. Postponement of bidder qualification until time of bid evaluation entails all of the problems associated with advertisement.

The rule of reason should apply to this dilemma. If qualification is relatively simple—for example, low-risk work involved, or most bidders are well known to the owner—or the qualification process short, qualify prior to bidding and avoid problems later. If qualification standards are exceptionally high and selective, or a time-consuming and expensive survey is required, make a preliminary screening prior to bidding and conduct detailed qualification as part of the bid evaluation process, focusing on only those candidate(s) who appear to be successful. The needs of the project, particulars of the contract, and schedule considerations all enter into this decision. As far as confidentiality goes, it is suggested that each qualified bidder (or others who inquire) be told only whether or not their company is on the bidder list, and not about the others.

REPRESENTATIVE CASES

1. Mr. X, an executive of an A–E firm's Western Division and project manager for a hydroelectric dam project, had just signed a contract award recommendation to let the general construction contract. By now the letter was in the mail advising his firm's client, the project owner, to choose contractor A for the work. Although contractor A was not the low bidder, his contract people had chosen them after an extensive analysis of the bids. Mr. X phoned his golf partner, corporate attorney for the A–E firm, and was told by the lawyer's secretary that he was in Chicago taking depositions on "the contractor A case"! After several frantic phone calls, Mr. X finally ascertained that contractor A and the A–E firm had been entwined in a bloody legal battle for years—with contractor A brutalizing them in the courts and press over a previous dam project for their eastern division. Mr. X was last seen chasing the letter carrier as she left the A–E firm's headquarters.

2. The contract administrator for a grass-roots manufacturing plant mailed RFPs to five contractors for the interior architectural work. All five had bid on the previous project some eight years ago and time was short, so he allowed only two weeks for bidding. On the bid due date, no bids were received. The contract administrator called all five bidders and received the following responses:

Bidder 1: Has since gone out of business

Bidder 2: Owner and general manager "out of the office" (spending four years in the federal penitentiary for conspiracy and fraud)

Bidder 3: Acquired two years ago by a huge conglomerate, which decided to drop the contracting "product line"

Bidder 4: Decided not to renew contractor's license in that state due to infrequent business there

Bidder 5: Decided not to bid "partially because we are in receivership and partially because you threw us off your last project"

3. In a great hurry to qualify additional bidders, a contract manager cannot get past the lowest clerk who answers the telephone at a contractor's bank. He refuses to divulge "confidential information" concerning contractor A's financial status. Frustrated and pressured, the administrator calls the chief executive officer of contractor A and tells him that he will not be asked to bid unless this information is provided by the bank. The contract manager finishes the call and walks across the office for a cup of coffee. When he returns there are three urgent phone message slips from the bank president. He returns the call, during which he is offered a complete set of financial statements on contractor A and invited to dinner at the bank president's private club to receive them.

CHAPTER 7

ISSUING REQUESTS FOR PROPOSALS

Once qualified bidders are selected for a particular contract, and the technical specifications, drawings, and other bid information have been prepared, you are ready to prepare a request for proposal (bid solicitation) and transmit it to the prospective bidders. Requests for proposals (RFPs) are composed of the following documents:

Invitation (instruction to bidders)
Proposal forms
Agreement
General conditions
Special conditions
Technical specifications
Drawings

During the initial document preparation period for the project (see Chapter 5), the format for all these documents is established, and certain ones are not altered for each RFP (such as the general conditions). The RFP development process, then, entails the completion of the remaining documents with variable information pertaining to the particular contract at hand. Each of these documents is discussed in the following sections.

The exhibits for this chapter appear on pages 98–118.

BIDDING DOCUMENTS

Invitation

This letter informs the bidders of the requirements for submitting bids, so particular information is included, such as the bid due date and time, number of copies of bids expected, project name and location, name and identification of contract, and date of prebid meeting, site visit, and so on. A sample letter is shown as Exhibit 2. It is helpful to alert bidders to any unusual bid requirements in this letter and to ask them to acknowledge receipt of the RFP as well as to return notification of their intention to bid (or not) to the sender. Some owners include a preaddressed postcard for this purpose. In order to identify the returning bids, preaddressed mailing envelopes or address labels can also be included.

Proposal Form

This document is prepared by the owner and completed by the bidders. The intention of a well-defined and well-structured RFP is that the proposal form is the only document altered (completed) by bidders. The proposal should be designed parallel to the agreement so that bidder-submitted data can easily be transposed to the agreement once an award is made. The proposal form is then filed, but not made part of the contract.

Since the proposal form is to contain the bidders' prices, its pricing section should be carefully prepared to solicit the intended contractual strategy and pricing alternatives. This usually constitutes the greatest effort of the proposal form preparation: designing the pricing section. It will be affected by the type of pricing selected (lump sum, unit prices, cost plus, and so on) and should, in all cases, provide for flexible pricing of any envisioned changes.

A sample proposal form for a lump-sum contract is included as Exhibit 3. note that it provides for preestablished unit prices for additions and deletions and requires the bidder to enter proposed markups for any cost-plus changes. Should the base contract be cost reimbursable, the form would be markedly different, as it would identify the allowable costs in great detail. This information is not usually required for a lump-sum contract.

Structuring Pricing Data. In soliciting pricing information, the natural tendency is to ask for only that which is necessary to select a low bidder. In the case of a lump-sum contract, for example, this may be only one dollar amount. Asking the bidders to show the components of this price (broken down by measurable, objective increments, in addition to the gross amount) will be helpful for a number of reasons:

1. Measurement of performance and determination of progress payments will be based on an objective assessment of actual accomplishment

rather than on subjective percentages of the total (progress billings and payments are discussed further in Chapter 12).

2. Unbalanced bids, such as front-end loading, will be surfaced.
3. Bidder error or misunderstanding as to scope of work will be isolated.
4. Baseline costs for additional or deleted segments of work will be established.
5. Historical, real-world cost data can be used to refine detailed owner project estimates for this and other projects.

In many cases, conflicts arise when the cost data are structured by one of the project management participants without regard for the needs of the other. Here are some typical situations:

Project cost estimates are prepared at a conceptual level or by element of cost orientation such that it is difficult to reconcile them with actual contract or subcontract line items.

Pricing structures in the RFP issued by the contract manager do not request cost data from bidders in a manner that facilitates calculation of "life-cycle" costs by the cost engineer.

RFP price sheets cause the cost engineer to reestimate the contract for the fair price estimate (fair price estimates are discussed in Chapter 9).

Contract line items are structured to match the way the project was estimated rather than the way the work will be performed. This makes it difficult to determine the value of work completed for progress payments.

A certain amount of give and take will be required, and the process of cost-related input to the bidding documents is usually an iterative exercise. But when it is successfully accomplished, all participants should be satisfied. The contract manager will have bids and eventual contract pricing items structured in a way that aids bid evaluation and contract administration, and the cost engineer or estimator, or accountant, will be able to update his or her estimating database and to report actual and projected costs throughout the project.

Additional Submittals. The proposal form should also list the required additional submittals and provide for the bidder's signature. It then becomes, in effect, an offer on the part of the bidder to perform the work, as defined, for the given price. Once accepted by the owner, a contract is formed.

When different options are solicited, or bidders are asked to modify or further define technical requirements, care should be taken to specify these explicitly in the RFP-and provisions should be made in the proposal form for each base bid, option, alternate, and details of bidder-provided technical or nonprice information. For example, if a performance specification is used, bidders can be encouraged to propose any number of alternatives that meet the standards given. In order to evaluate these alternatives, the proposal form should be explicit and easy to understand. When technical submittals are needed to evaluate these properly, their format, content, and specific data needed should be specified.

It is a good idea to include several extra copies of the proposal form with the RFP (say, five). This is convenient for bidders but also allows the owner to ask for more than one completed proposal to ease its evaluation task, especially if several internal groups will evaluate bids simultaneously (e.g., cost, technical, commercial). Sometimes owners specify that pricing information be contained on some copies and omitted from others. This allows noncost evaluations to be made without stringent bid security measures.

SPECIMEN CONTRACT DOCUMENTS

The documents described in the following subsections are included with an RFP as specimens of the anticipated contract documents.

Agreement

This document, along with the remainder of the package, is sent to inform the bidders of the contents of the final contract once a bid is accepted. Upon award, it is completed and signed and forms the heart of the contract documents. A sample is shown as Exhibit 4.

General Conditions

These should require no specific alteration for the RFP. A sample table of contents for a set of general conditions is given in Exhibit 5.

Special Conditions

Contract-specific commercial terms are included here. As shown in Exhibit 6, the special conditions describe the scope of work, requirements for contractor's employees, material, equipment and labor to be furnished by the contractor, owner, and others, schedule and general sequence of work required; list the applicable drawings; and define how work will be measured and payment made.

Technical Specifications

Before the RFP is prepared, the technical specifications (specs) must be reviewed to ensure that they are complete, that the scope of work descriptions contained in the other documents agree with the specs, and that no commercial or legal terms or conditions are included in them.

Drawings

Usually at least one copy of each drawing required to perform the work is included in the RFP packages. Some owners include two or more copies for the bidders' convenience. The drawings should be listed as to title, number, revision status, and date of issue in the RFP (as suggested here in the special conditions) so the bidders can check to ensure that they have all proper versions of the drawings. It is not unusual for the wrong drawing to be mailed to bidders or for some drawings to be omitted, illegible, or outdated. The risk of error is intensified under schedule pressures and with large quantities of drawings and fast-tracked engineering effort.

ISSUING THE REQUEST FOR PROPOSAL

Once prepared and reviewed by all affected parties, (engineering, estimating, project management), the RFPs are ready to be mailed or submitted to potential bidders. Written documents are usually bound in a booklet or binder; drawings are printed and mailed. Several copies are sent to each bidder, so they can send some to subbidders. Some owners issue reproducible drawings, such as sepia copies, to bidders for this reason. The contract manager should contact each bidder within a few days to ensure that it has received the RFP. When fair price estimates are used (see Chapter 9), the estimator or cost engineer should receive the identical package as sent to bidders and be allowed the identical time to complete his or her "bid." Once the RFPs are issued ("on the street"), requests may come to the owner for additional copies of documents or drawings. It is up to the owner to set a policy on reproduction and mailing costs. Often, these requests come from subbidders who will be quoting a portion of the contract work to one or more eventual bidders. A good control policy is to respond only to *bidder* requests, not to subbidder needs. If subbidders need copies (or information), they should be instructed to communicate through bidders. This brings up the question of whether subbidders should be told who the prime bidders are. It is an owner's prerogative to do so when bidding is by invitation, but to avoid problems, do not do this unless you have a good reason.

Some owners require all RFP documents to be returned with the bid, or otherwise if the bidder declines to bid. Other owners will establish RFP "viewing rooms" at various locations where interested subbidders may view or copy the RFP documents. This is unusual in invitation-only circumstances.

REPRESENTATIVE CASES

1. A contractor had been selected by the owner of a proposed $200 million office tower complex. The design was fairly well defined, but recent fluctuations in the market price of major building materials in addition to the known penchant of the owner's president to change his mind concerning interior design led the owner to negotiate a guaranteed maximum price with the potential contractor. Since only one contract was to be awarded for the project, the task of designing the contract and bidding documents and that of preparing a specific RFP were combined. The intention was for the contractor to "bid" the guaranteed maximum amount using the RFP documents.

Unfortunately, in designing the proposal form, no thought was given to the eventual task of evaluating the contractor's guaranteed maximum amount. When the bid was received, the owner found that the contractor had broken his bid amount down according to the project's major structures: office tower, service building, parking garage, pedestrian mall shops, and so on. The project estimate, however, had been prepared by the architectural consultant according to material quantities and elements of costs: structural steel, concrete, elevators, labor, concrete formwork, roofing, HVAC systems, and so on. Needless to say, meaningful comparisons between the two price structures were impossible. And, as luck would have it, the bottom-line guaranteed maximum price exceeded the architect's estimate by 40%. The owner, under schedule pressure, was forced to either:

1. Accept the contractor's bid and proceed on faith that it was reasonable.

2. Decrease the size or quality of the project to meet the budget established as a result of the architect's estimate. But where to cut?

3. Reestimate the project according to the contractor's bid format. This would make the evaluation feasible, but the resulting contract price structure would make objective measurement of progress and corresponding progress payments difficult.

4. Rebid the project after defining an objective work breakdown structure, reestimating the project according to this structure, and asking the contractor to conform his bid to such a structure.

For the benefit of all concerned, particularly its own, the owner chose the proper course of action: option 4.

2. A construction management firm decided that since bid drawings change so frequently during the bid cycle, the use of a drawing list in the special conditions would only add another section needing revisions through addenda. A lump-sum contract served as the basis for reversing this policy. Of 362 drawings issued for bid, with revisions occurring daily, it was inevitable that a contractor would receive and base its bid on some superseded drawings. In this case, the low bidder and eventual contractor did not include 19 large electrical manholes, the substitution of nickel–chromium pipe for carbon steel pipe, the addtion of concrete pads for buried storage tanks, and

several other expensive changes brought about through drawing revisions as the RFP was being prepared. All other bidders received and noted the drawing changes and were therefore higher in bidding. Resulting change orders to the eventual contractor had the cumulative effect of making it what would have been the highest bidder. Had it received the appropriate bid drawings, this problem would have been avoided.

3. General conditions used for cost-plus contracts issued by a consulting firm called for reimbursement of "all reasonable overhead, general and administrative expenses providing such are substantiated by purchase or rental invoices, payroll records, stock depletion records, and the like." This broad and general language was the cause of delight and sebsequent abuse for a general contractor hired to design and build an industrial plant for the consultant's client. Absent precise conditions specifying allowable and allocatable expenses for the project, the general contractor passed through expenses that normally would be denied. Only a small sampling of such expenses is given here to demonstrate the need for precise definition of reimbursable expenses in every cost-plus contract:

New personal computer and software

New radios and transmitters

New vehicles for all management personnel

New ambulance

New CAD-CAM equipment

Complete overhauls of cranes, trucks, compressors, and other major equipment

Purchase of 30 welding machines

Tuition and expense reimbursement for attendance at 27 seminars by contractor personnel

Three-year lease of two copy machines

New hard hats and safety equipment

One-year personal services consulting contract for a retired member of the contractor's board of directors

Travel expenses for project manager and contractor's president to Florida, Hawaii, Bermuda, and London to meet vendors and/or visit similar projects under construction

And so on. . . .

EXHIBIT 2. Sample Invitation to Bid

Date

(Bidder's name and address)

(Owner's name, project name, description of work, and contract number)

You are hereby invited to submit your proposal for subject work located (site location) for (owner's name).

Proposals are due before (bid due date, time, and time zone) in the offices of _____ at the following address:

> _____
> _____
> _____
>
> ATTN: _____

Preparation of proposals shall be at the Bidder's expense, in accordance with the enclosed Specimen Contract Documents, and on the Proposal Forms provided herewith. Interlineations or alterations to these forms will not be acceptable. Bidders shall list any exceptions or clarifications on a separate sheet of paper, bearing their letterhead, with suitable references to the specific items as addressed in the Contract Documents.

Proposals shall be submitted in the following manner.

1   signed original (priced)
3   signed copies (priced)
3   unsigned copies (complete except for price listing)

No proposal security is required.

Requests for interpretation of the Request for Proposal documents shall be submitted in writing to Ms. _____ at the above address. All changes to these documents will be made by the Owner in the form of written addenda issued to all Bidders.

Bidders are advised (or required for consideration of their proposals) to visit the job site to acquaint themselves thoroughly with all conditions which may affect their proposals or performance of the work. Arrangements for site visits shall be made with the Owner through Mr. _____ at _____ .

EXHIBIT 2. *(continued)*

A prebid meeting will be held on (date) at (time) at (location). Failure of a Bidder to attend may be cause for rejection of his proposal.

Should additional copies of the Request for Proposal documents be required, contact _____. Requests from noninvited contractors, subcontractors, or other third parties will not be honored.

Bidders are hereby informed that the contract between the Owner and the successful Bidder will be the legally controlling document and that all communications, verbal or written, made prior to or in addition to the Contract Documents signed by both parties will be abrogated, withdrawn, and of no legal force or effect.

The Owner reserves the right to open Proposals privately and unannounced. The Owner does not obligate itself to accept the lowest or any other bid and reserves the right to reject any and all bids.

Kindly inform _____ by (date) as to your intention to bid and to attend the scheduled prebid meeting.

Very truly yours,

_____
(signature)

EXHIBIT 3. Sample Proposal

PROPOSAL
- FOR
(Description of work to be covered
by the contract and contract number)
FOR
(Project description)
OF
(Owner's name and location)

The undersigned Bidder herewith proposes to do all the work, perform all the services, and furnish all the materials, except the specific materials and services to be furnished by the Owner, required for (description of work to be covered by the contract) for the (project description) of (owner's name); the work site being located at (location of work). The scope of work is set forth in the attached specimen Contract Documents labeled Contract (contract number), and all addenda which may be issued thereto.

The Bidder agrees that in case the Contract is awarded to it, it will begin actual work on _____ or within _____ calendar days after receipt of notice to proceed. The Bidder has attached to this Proposal a schedule for the work and agrees to maintain a rate of progress set forth in said schedule and that it will complete the work on or before the dates of completion required by Contract Document (contract number).

The Bidder proposes to complete all work as defined in Contract Document (contract number) for the lump-sum amount of _____dollars ($_____), which is composed of _____ dollars ($ _____) material cost and _____ dollars ($ _____) labor cost.

A breakdown of the Contract Price is as follows:

I. Lump Sum

Item no.	Description	Material	Labor
1		$	$
2		$	$
3		$	$
	Total material cost	$	
	Total labor cost		$
	Total lump-sum cost	$	

II. Additions and Deletions

The following unit prices will apply for additions or deletions to the amounts included under Item I above:

Item no.	Description	Unit of measurement	Unit price	
			Material	Labor
1			$	$
2			$	$
3			$	$

III. Unit Prices (if used)

The following unit prices will apply to work not included under Item I above:

Item no.	Description	Estimated quantity	Unit of measurement	Unit price	
				Material	Labor
1				$	$
2				$	$
3				$	$
4				$	$

Should the Bidder, as Contractor, be requested to perform any additional work on a cost-plus basis, the proposed fees, as defined in Contract Document (number), General Conditions, Article (changes clause), are as follows:

_____ % for direct labor

_____ % for material

_____ % for subcontracted work

When performing cost-plus work, craft labor wages including fringes and benefits shall be in accordance with prevailing local rates. The rates shall be submitted in compliance with the attached Sample Wage Rate Breakdown to be completed by the Bidder and submitted with the Bid.

Equipment rates for additional work shall be discounted at _____ % of those shown in the current A.E.D. Rental List or as shown on the attached Equipment Rental Schedule to be completed by the Bidder and submitted with his Bid. Rental lists or schedules shall show rates for monthly, weekly, and daily rentals, and those rates producing the most economical result for the Owner shall apply, in accordance with the actual circumstances of the case in each instance.

The Bidder proposes to credit the Owner the following amounts to be deducted from the total Contract Price, in the event the Owner provides the stated material or services:

1. (List of material or services owner may provide $(   )
2. at its option, over and above those listed in
    the specimen contract documents) $(   )
3. $(   )

(continued on p. 102)

EXHIBIT 3. *(continued)*

The Bidder requires the following temporary facilities:

1. Site space requirements for field office and storage   _____ sq. ft.
2. Electric power requirements   _____ kVA

The Bidder agrees that this Proposal constitutes a firm offer to the Owner and cannot be withdrawn for ninety (90) calendar days after the due date. The Bidder also agrees that the prices listed in the Schedule of Prices are firm and shall remain so throughout the performance of the work.

The Bidder proposes to pay all federal, state and local income, gross receipts, and franchise taxes due and payable by Contractor in connection with all work provided for in the Contract Documents, to make any and all payroll deductions required by law, and to hold the Owner and the Engineer harmless from any liability on account of any such taxes or withholdings.

The Bidder agrees to perform any additional work as requested under the terms and conditions of the Contract Documents.

The Bidder proposes to subcontract specific portions of the work to the following subcontractors:

Subcontract work	Subcontractor's name and address	Estimated % of total contract amount

No work other than that listed above shall be subcontracted unless approved by the Owner. Bidder agrees that, if awarded the Contract, it will submit to the Owner copies of all subcontracts and necessary insurance information, in accordance with the General Conditions, before each subcontractor starts work.

The Bidder shall attach the following information to its Proposal.

1. Proposed job superintendent's résumé, including a listing of recent jobs of this nature, areas of responsibility, duration, and job references including names and telephone numbers
2. A simplified Critical Path Method precedent diagram or a bar chart indicating Bidder's suggested start and completion dates for the principal sections of the work
3. Labor loading curves for the various crafts under its direction
4. Equipment rental schedule
5. Sample wage rate breakdown
6. (List any other submittals required with the proposal, such as a list of foreign materials, organization chart with names of proposed field staff, proposed delivery schedules, and any submittals required with the bid by the technical specifications.)

The Bidder proposes to furnish to the Owner, if requested, within ten (10) days after execution of the Contract, a surety company bond, in the form provided herein, for the faithful performance of the Contract including the payment of all labor and indebtedness to equipment rental firms, material suppliers, and any other vendors or subcontractors to whom the Bidder may be obligated in respect to the performance of the Contract. The penalty of the performance bond shall be one-hundred percent (100%) of the Contract Price. Cost of such bond will be borne by the Owner and shall not be included in the prices listed in the Schedule of Work.

Proposed surety is:

Full Name _____

Address _____

A Corporation organized under the laws of the state of _____ and duly authorized to transact business in the state of _____ . Cost of premium to be borne by Owner is $ _____ per $1,000 of bond.

Any jurisdictional dispute that may arise in connection with work performed pursuant to the terms of the Contract shall be settled in accordance with the Plan for the Settlement of Jurisdictional Disputes in the Construction Industry and any decision by the Impartial Jurisdictional Disputes Board shall be final and binding.

In performing the work hereunder, the Bidder, as Contractor, proposes to adopt wages, working conditions, and other employment policies that meet with the approval of the Owner and, where applicable, shall comply with the wages and working conditions established in the agreements with local unions affiliated with the Building and Construction Trades Department (AFL-CIO) having jurisdiction over the specified work, provided that the Bidder, as Contractor, will not be required to violate any applicable federal, state, or local codes, regulations, or statutory provisions.

The undersigned Bidder declares that it or its agent has carefully examined the location of the proposed work, the Contract Documents, and the Drawings for the Contract, is informed as to local conditions affecting the proposed work, has completed the requested attached information, and understands that the Owner reserves the right to reject any or all bids.

The Bidder affirms that, in making this Proposal, neither the Bidder, any company it represents, nor anyone on its or its company's behalf, has directly or indirectly entered into any combination, collusion, undertaking, or agreement with any other bidder or bidders to maintain the prices of said work or any compact to prevent any other bidder or bidders from bidding on the work described in this Proposal. The Bidder further affirms that this Proposal is made without regard or reference to any other bidder and that it is made without any agreement, understanding, or combination with any person or persons concerning such bidding in any way or manner whatsoever.

(continued on p. 104)

EXHIBIT 3. *(continued)*

The Bidder is (please indicate below):

An Individual _____

A Corporation of the state of _____

A Partnership consisting of _____

A Joint Venture comprised of _____

Full company name _____

By _____

Title _____

Date _____

Business address _____
_____

Telephone number _____

(State) Contractor license number _____

The Bidder states that the following individuals are authorized signatories of contracts binding on the Bidder for not less than the total Contract Price as stated herein:

Name	Title
_____	_____
_____	_____
_____	_____

The Bidder acknowledges receipt and consideration of the following Addenda to the Request for Proposal:

Addendum no.	Date issued
_____	_____
_____	_____
_____	_____
_____	_____

Sample Wage Rate Breakdown (to be completed by Bidder)

Contractor: _____

Craft: _____

Classification: _____

Union local no. and address: _____
_____

Description	Rate (per hour) Straight time	Rate (per hour) premium time	Rate (per hour) Shift time
(1) Base wage	$	$	$
(2) Fringes and benefits (list individually contributions to funds and other fringe benefits required under terms of collective bargaining agreements or company policy)			
a.	$	$	$
b.	$	$	$
c.	$	$	$
d.	$	$	$
e.	etc.	$	$
(3) Payroll taxes and unemployment insurance (list individually Social Security taxes, Unemployment Insurance, Workers' Compensation, and so on, including amounts and percentages)			
a. @ %	$	$	$
b. @ %	$	$	$
c. @ %	$	$	$
d. @ %	$	$	$
e. @ %	$	$	$
(4) Markup for overhead and profit (per contract) %	$	Not applicable	
(5) Total wage rate (Total of items 1 through 4) per hour.	$	$	$

Note: Separate sheets shall be submitted for each classification (such as apprentice, journeyman, and foreman).

Equipment Rental Schedule (to be completed by Bidder)

The following rates include fuel, lubrication, and all maintenance but without operator. No overtime premium will be paid on equipment. The Owner reserves the right to select the rental period that proves to it to be the most economical.

(continued on p. 106)

EXHIBIT 3. *(continued)*

Equipment	Description	Rate per hour	Rate per week	Rate per month
_____	_____	_____	_____	_____
_____	_____	_____	_____	_____
_____	_____	_____	_____	_____
_____	_____	_____	_____	_____
_____	_____	_____	_____	_____

EXHIBIT 4.   Sample Agreement

AGREEMENT
FOR
(description of the work)
FOR
(Owner's name)
(project description)

This AGREEMENT, made and entered into (by letter of intent as of the, or) this _____ day of _____ in the year One Thousand Nine Hundred and _____(19____), by and between (owner's name) hereinafter called the Owner, and (contractor's name), whose office is located at (contractor's business address), hereinafter called the Contractor,

WITNESSETH that the parties to these presents, each in consideration of the undertakings, promises, and agreements on the part of the other herein contained, have undertaken, promised and agreed, and do hereby undertake, promise and agree, the Owner for itself, its successors and assigns, and the Contractor for himself, his successors and assigns, as follows:

FIRST—The Contractor agrees, at its own sole cost and expense, to perform all the labor and services, and to furnish all the materials and permanent equipment except as herein otherwise specifically set forth, and to furnish all the construction plant and equipment necessary to complete in a good, substantial, and approved manner within the time hereinafter specified and in accordance with the terms, conditions and provisions of this Contract, and of the instructions, orders and directions of the Owner made in accordance with this Contract, the (description of the work), at the Owner's site located (project location).

The above works are more completely described in the General Conditions, Special Conditions, Technical Specifications, and the Drawings which are part hereof.

The Contractor agrees to protect the said works from damage by the elements or otherwise until completed and delivered in accordance with the Contract Documents.

SECOND—The Owner agrees to furnish to the Contractor the materials and services as provided for under Article (number) of the Special Conditions.

THIRD—The Contractor agrees to comply with the schedule as provided for under Article (number) of the Special Conditions.

FOURTH—The Owner agrees to pay and the Contractor agrees to accept as full compensation, satisfaction, and discharge for all work performed and all materials installed, and for all costs and expenses incurred, damages sustained and for each and every matter, thing or act performed, furnished or suffered in the full and complete performance and completion of the work of this Contract in accordance with the terms, conditions and provisions thereof, and of the

(continued on p. 108)

EXHIBIT 4. *(continued)*

instructions, ordered and directions of the Engineer thereunder, except changes or additions covered by Contract Change Orders, which shall be paid for as provided in the General Conditions, a lump sum equal to _____ Dollars ($).

A Breakdown of the Contract Price is as follows:

Lump-Sum Schedule

		Amount	
Item no.	Description	Material	Labor
1	(breakdown of lump-sum items, as shown in the schedule of prices, Item I of the proposal)	$	$
2		$	$
3		$	$
4		$	$
5		$	$
6		$	$
	Total material price	$	$
	Total labor price		$
	Total contract price	$	

For additions and deletions to the work covered in the Lump-Sum Schedule above, the Contract Price will be measured and derived from the amount of actual work performed and materials installed, as determined by the Engineer under each item of the following schedule, multiplied by the unit price applicable to each such item as set forth in the following schedule:

Unit Price Schedule for Additions and Deletions

			Unit price	
Item no.	Description	Unit of measure	Material	Labor
1	(unit price schedule for additions and deletions as shown in Item II of the schedule of prices in the proposal)		$	$
2			$	$
3			$	$
4			$	$
5			$	$

For work that is included in the scope of work under this Contract, but that is not covered by the Lump-Sum Schedule above, the following Unit Price Schedule shall apply. Payment will be a sum to be measured and derived from the amount of actual work performed and materials installed, as determined by the Engineer, multiplied by the unit price applicable to each item as set forth in the following schedule:

Unit Price Schedule

Item no.	Description	Unit of measure	Unit price	
			Material	Labor
1	(unit price schedule as shown in Item III of the schedule of prices in the proposal. Note that estimated quantities are not given in the contract agreement but serve only as a basis for bidding and are therefore withdrawn after receipt of proposals.)		$	$
2			$	$
3			$	$

The measurement and payment of all items included in the above schedules shall be as prescribed in Article (number) "Measurement and Payment" of the Special Conditions.

Should the Contractor be requested to perform any additional work for which it will be compensated in a cost-plus manner as specified in Article (number) of the General Conditions, the Contractor's fees for such work shall be as set forth in the following schedule. Equipment rental rates paid shall be as shown on the attached Equipment Rental Schedule, or shall be discounted at _____ % of the current A.E.D. Rental List if not shown on the attached Schedule. Equipment rental reimbursement will be on the basis of monthly, weekly, or daily rates as actually contracted for by the Contractor with the equipment lessor.

Additional Work Pricing Schedule

(Insert the pricing schedule for additional work as shown in the proposal.)

FIFTH—All the provisions contained in this Agreement, General Conditions, Special Conditions, Technical Specifications, and Drawings and those items included by reference shall be deemed terms and conditions of the Contract as if fully embodied herein.

(continued on p. 110)

EXHIBIT 4. *(continued)*

These Contract Documents are complementary, and what is called for by one of them shall be as binding as if called for by all.

SIXTH—Without limiting any of the other obligations or liabilities of the Contractor, the Contractor shall adhere to the General Conditions and provide and maintain insurance in accordance with the provisions set forth therein until the work is completed and accepted by the Owner.

The contractor shall before commencing work on this Contract, deliver to (name) at (address) two copies of Certificates of Insurance, completed by his insurance carrier or agent certifying that minimum insurance coverages required by the Contract Documents are in effect.

SEVENTH—The Contractor agrees to pay all federal, state, and local income, gross receipts and franchise taxes and taxes of every other nature due and payable by the Contractor in connection with all the work provided for in the Contract, and to make any and all payroll deductions required by law, and to hold the Owner and its agents harmless from any liability on account of any such taxes or withholdings.

(Insert specific provisions for local sales and/or use taxes as applicable.)

EIGHTH—Any jurisdictional dispute that may arise in connection with work performed pursuant to the terms of the Contract shall be settled in accordance with the Plan for the Settlement of Jurisdictional Disputes in the Construction Industry and any decision by the Impartial Jurisdictional Disputes Board shall be final and binding.

In performing the work hereunder, the Contractor shall adopt wages, working conditions and other employment policies that meet with the approval of the Owner, and, where applicable, shall comply with wages and working conditions established in the agreements with local unions affiliated with the Building and Construction Trades Department (AFL –CIO) having jurisdiction over the specified work provided that the Contractor will not be required to violate any applicable federal, state, or local codes, regulations, or statutory provisions. (Article Eight is applicable when union labor is required by a project agreement or other understandings and may not always be applicable.)

NINTH—The Contractor, within ten (10) days after the execution of this Contract, shall furnish to the Owner a surety company bond in the form hereto attached with such surety as shall be satisfactory to the Owner, conditioned in the amount of one hundred percent (100%) of the Contract Price for the faithful performance of this Contract.

(delete above article if bond is not required)

IN WITNESS WHEREOF, the Parties hereto have duly executed this Agreement the day and year first written.

_____(owner's name)_____

By _____
                    Name

_____
                    Title

Witness: _____

_____(contractor's name)_____

By _____
                    Name

_____
                    Title

Witness: _____

_____

_____

GENERAL CONDITIONS
FOR
(owner's name)
(project description)
INDEX

Page

1.0. Definitions of terms used ....

2.0. Time and order of completion ....

3.0. No waiver of obligations ....

4.0. Owner inspection and right of access ....

5.0. Contractor warranties and correction of defective work ....

6.0. Discontinuance of work by owner ....

7.0. Contract parts and owner's decisions ....

8.0. Changes in the work ....

9.0. Provisional acceptance of portions of the work shall not constitute a waiver ....

10.0. Increase of contractor working force and equipment ....

11.0. Construction lines and grades ....

12.0. Materials and equipment to be used ....

13.0. Contractor's address ....

14.0. Progress estimates and payments ....

15.0. Final payment ....

16.0. Personal attention of contractor ....

17.0. No claim because actual quantities differ from estimates ....

18.0. Subcontracts ....

19.0. Infringement of patents ....

20.0. Work performed at contractor's risk ....

21.0. Contractor informed as to conditions ....

22.0. Assignment of contract ....

23.0. Right of termination by the owner ....

24.0. Regulations and permits ....

25.0. Lands for construction purposes ....

26.0. Use of explosives ....

27.0. Protection of permanent and temporary facilities ....

28.0. Intoxicants, narcotics, and firearms ....

29.0. Character of employees ....

30.0. Contractor agents, superintendents, and foremen ....

31.0. Suspension of work if contract is violated ....

32.0. Bonding requirements ....

33.0. Removal of equipment and material ....

34.0. Indebtedness and liens ....

35.0. Indemnification and insurance ....

36.0. Collateral work ....

37.0. Right to temporarily suspend work by the owner ....

38.0. Workers to be used . . . .
39.0. Final dressing and cleaning up . . . .
40.0. Independent contractor . . . .
41.0. Fire protection . . . .
42.0. Requirements for health and safety . . . .
43.0. Equal opportunity . . . .
44.0. No other understandings . . . .

EXHIBIT 6.   Sample Special Conditions

SPECIAL CONDITIONS
FOR
(description of work and contract number)
FOR
(project description)
OF
(owner's name)

TABLE OF CONTENTS

Page
1.0. Scope of work..
2.0. Project job rules..
3.0. Furnished by contractor.......................................
4.0. Furnished by owner..
5.0. Schedule and sequence of work ....
6.0. Drawings...
7.0. Measurement and payment for work..............................

1.0. Scope of work

1.1. These Special Conditions, together with the Agreement, General Con-
      ditions, Technical Specifications, and Drawings cover the work of (de-
      scription of work, project description, and owner's name and site location).

1.2   (Further descriptions of work scope, such as buildings, areas, systems,
      limits, and work specifically excluded from this contractor's scope.)

2.0. Project job rules

2.1   (Insert or attach job requirements such as parking areas, brassing, safe-
      ty helmets with identification colors, project identification badges, off-
      limits areas, and access to existing facilities)

3.0. Furnished by contractor

3.1. The Contractor shall furnish all material, supplies, labor, services, su-
      pervision, tools, plant, apparatus, conveyances, equipment, temporary
      buildings, scaffolding, transportation, overhead, and incidental expense
      for accomplishing the work covered by this Contract, except the materials
      and services specifically named herein to be provided by the Owner.

3.2. Should the Contractor require additional facilities beyond those provided
      by the Owner, the Contractor shall provide its own additional facilities,
      such as for storage of materials, offices, change houses, or shops in the
      area allotted by the Owner.

3.3. The Contractor shall furnish the Owner with a breakdown of the Contract Price (based upon the actual cost of materials furnished and services performed) as may be required by the Owner to establish the cost of individual units of property and integration into his code of accounts. This requirement shall be satisfied prior to preparation of the initial progress estimate for payment under the terms of the Contract, unless otherwise specified by the Owner.

3.4. (The remaining articles under this section identify specific contractor-furnished items and may include some or all of the following:

   Temporary power and water
   Sanitary facilities
   Layout of the work
   As-built drawings (detailed instructions on this subject are required)
   Reports, completed forms
   First-aid facilities
   Fencing, security

4.0. Furnished by Owner

4.1. The Owner will furnish, either directly or through others, without cost to the Contractor, the following materials, equipment, and services under the conditions described below.

   (Insert owner-furnished items. A short description of location and conditions pertaining to each item should also be presented here. The following are examples of such items.)

   1. Project identification badges indicating Contract number and employee number.
   2. Space for Contractor's temporary building, at site, in area designated by the Engineer.
   3. Water for drinking purposes, at valve at designated location at plant site. Storage or cooling of drinking water will not be provided.
   4. Construction power 460 volts, 3 phase, 60 hertz at designated temporary power panels. The Contractor shall provide and maintain the required connections and cable from load center to its equipment. Power will not be furnished for heating or cooling Contractor's construction buildings.
   5. Workers' sanitary facilities, on site, to be shared with other contractors.
   6. (List specific material, equipment, or services related to the work under this contract but supplied or performed by others.)

4.2. All questions pertaining to and requirements for the above items shall be directed to the owner.

(continued on p. 116)

EXHIBIT 6. *(continued)*

4.3.* The Contractor shall receive materials furnished by the Owner on delivery conveyances at the job site, unless materials have been delivered prior to the award of this Contract, in which case Contractor shall receive materials on the ground as they occur or from Owner's storeroom on the site.

4.4.* The Contractor shall receive, unload, check, and store the materials and equipment furnished by the Owner, in cooperation with the Owner, and shall furnish the latter with a materials receiving report duly signed by the Contractor's authorized representative signifying that the materials and equipment were received in good order. Shortages, errors or damages shall be called to the attention of the Owner and acknowledgment with its signature so noted on the Owner's receiving report, or carrier's freight bill or delivery form, and a shortage form or damage report shall be completed in cooperation with the Owner. Contractor's responsibilities in accordance with the provision of Article 4.0 of these Special Conditions shall commence at the time of award of this contract. If the Contractor is unable or refuses to perform the receiving, unloading, storing, and maintenance as required herein, the Owner may perform such duties of the Contractor, either by itself or by others, and backcharge the Contractor for all costs associated therewith.

4.5.* After the Contractor has accepted materials and equipment furnished by the Owner as being in good condition and correct quantity at time of delivery, it shall be responsible for their storage, maintenance and safety from loss or damage of any nature until the finished work and/or surplus materials are accounted for and accepted by the Owner.

4.6.* The Contractor shall pay all demurrage charges on transportation equipment containing materials furnished to the Contractor by the Owner.

4.7.* The Contractor shall pay for all damage to Owner-furnished equipment or materials accruing as a result of Contractor's unloading or transportation operations.

5.0. Schedule and sequence of work

5.1. The Contractor shall be mobilized on site on or before _____ and shall commence and proceed with the performance of the work with speed and diligence so as to ensure its completion on or before _____ .

5.2. The Contractor shall commence and complete the principal sections of the work in accordance with the following schedule:

Description	Start work	Complete work
1. (list the principal sections of the work, such as		
2. structures, systems, floor elevations, areas, and		
3. equipment, with the appropriate start and end		
4. dates)		

* These articles apply when the Owner is furnishing the Contractor materials or equipment.

(if scheduling information is required by the contractor for auxiliary or support services or systems, such "target dates" should also be included.)

5.3. The Contractor shall assist the Owner, on a periodic basis, in preparing detailed schedules of Contractor's remaining work as performance progresses. The Owner will prepare an overall schedule for the total work incorporating the Contractor's detailed schedule. The Contractor's schedule will be revised when necessary to indicate any changes required by the Project conditions.

5.4. (If periodic progress or quantity-installed reports are required, they should be listed here.)

6.0. Drawings

6.1. It shall be the responsibility of the Contractor to notify the Owner, without delay, of any omissions, errors, or discrepancies the Contractor may discover in the Drawings. The Contractor shall in no case proceed without full resolution by the Owner of omissions, errors, or discrepancies.

6.2. A maximum of _____ copies of each required Drawing will be furnished to the Contractor without charge. Any additional copies will be charged to the Contractor at invoice cost.

6.3. Drawings furnished by the Contractor shall become Contract Drawings after acceptance by the Owner.

6.4. The following drawings are the "Drawings" that are referred to in various parts of this Contract and are considered as a part hereof:

Drawing no.	Revision	Date	Title

(List the drawings pertaining to this contract. Drawings from each engineering discipline should be grouped and identified as such, and "reference" drawings should be separately listed and identified. If a large number of drawings are used, they may be listed in a separate "List of Drawings" and inserted after the technical specifications, rather than listed here.)

7.0. Measurement and payment

7.1. Payment for Item I of the Schedule of Prices will be made on a lump-sum basis. Partial payments based on actual physical progress will be made in accordance with the performance payment schedule set forth in Article 7.8. The lump-sum payment, divided to state separately material cost and labor cost, shall include all costs of performing the work in accordance with the Contract Documents.

7.2. Payments for additions or deletions from the work included in Item I of the Schedule of Prices and occasioned by substantial changes in the

(continued on p. 118)

EXHIBIT 6. *(continued)*

Drawings will be made on a unit-price basis as defined in Item II. The quantity to be paid for, or credited, will be the net total of units added, or deleted, as determined from the Drawings or as determined from the estimated quantities if they are used in Item I of the Schedule of Prices. The unit prices shall cover all costs of performing the work in accordance with the Contract Documents.

7.3. The quantities to be paid for under Item III of the Schedule of Prices will be the total number of units of each item in place as shown on the Drawings. The unit prices shall cover all costs of performing the work in accordance with the Contract Documents.

7.4. General

  1. Material costs shall include all permanent and temporary material incorporated into the work.
  2. Labor costs shall include wages and employee benefits, both employer, union and government furnished and employer wage taxes, and also include the unloading, storing, transferring, installing, maintaining, documenting, identification, and testing of material and equipment, and the training, testing, and qualifying of craft personnel.

7.5. Lump Sum—Specific Items

(Include any required clarification of the work that is covered under the lump-sum prices for each item in the Schedule of Prices; for example, the lump-sum price for subgrade ductbanks shall also include required concrete, excavation, and backfilling.)

7.6. Additions and Deletions, and Unit Prices—Specific Items

(Include any required clarification of the work that is covered under the additions and deletions and the unit price sections for each item in the Schedule of Prices; for example, the unit price for cable tray shall also include all hangers, supports, anchors, spacers, minor field fitting of tray, temporary supports and hangers, and identification.)

7.7. (Any other methods of payment should be clarified here. Examples are: per diem for manufacturer's representative, including expenses, weekly or monthly payments for storage or maintenance, credit to the owner for optional owner-furnished services, and so on.)

7.8. (A definite progress payment schedule, including percentages of total payment for each completed operation, should be given for those sections of the work that involve several operations over long periods of time.)

CHAPTER 8

MANAGING THE BID CYCLE

The bid cycle begins with the issuance of RFPs to bidders and ends when the owner receives completed bids. During this period the owner is tempted to reduce its active role in the contract formation process. Quite often, owners feel that once RFPs have been released, their control objectives are minimal—that the initiative has been transferred to the bidders, that "it's in their hands now, there is little we can do until the bids are received and we begin our evaluation." This passive role is not recommended, for there are many actions the owner may take that will improve the quality of submitted bids, ease the evaluation and award efforts, and in fact reduce the price of expected bids. In other words, the bid cycle is no different than other formation events—it should be actively controlled, or managed, by the owner to protect its best interests.

Let's begin our discussion by considering the reasons why a bidder may price a proposal higher than expected. Common reasons for unreasonably high bid prices are:

1. *Complementary Bidding.* This occurs when the bidder is not interested in performing the work (for any number of reasons) yet does not want to offend the owner or otherwise to be prevented from bidding on future work—to be deleted from future bidder lists. In these cases, bidders sometimes inflate their proposed price to a range that they believe will prevent award. They hope to avoid being asked to perform the work yet are not forced to decline the bidding opportunity.

2. *Bidder Error.* The bidder may simply make mistakes when calculating bid prices.

3. *Bidder Misunderstanding of Scope of Work and/or the RFP Documents*. Clear and complete RFP's, addenda, prebid meetings, and site visits help reduce this possibility.

4. *Perceived Risk on the Part of the Bidder*. Any number of factors may cause the bidder to perceive real or imaginary areas of risk and to raise its bid price in order to provide for this risk. Depending on its understanding of the scope of work and project conditions, not to mention its bidding philosophy, it may unnecessarily pad the price to cover imagined risks.

5. *Bidder Uncertainty*. The bidder may be quite uncertain about the proposed work, the owner's objectives and management plans, or its own ability to perform as expected. To provide protection from uncertainty, it may increase his price beyond that which is justified. Or, in some cases, the owner and its representatives are so uncoordinated and incompetent that the contractor is smart to pad the price.

Each of these factors leads to bids priced higher than they would be under better conditions. The owner's major responsibility during the bid cycle is to eliminate these bidding factors or reduce their impact on bidder prices. This is done through a controlled program of information dissemination—by letting the bidders know as much as possible about the bidding conditions, eventual contract terms, and anticipated conditions once construction begins. Here's how this is done:

1. Answer or respond to bidder questions in a clear and timely manner so that identical information is available to all bidders
2. Prepare and issue addenda (changes to the RFP) to clarify any changes made from the time the RFP is released until bids are received
3. Extend or change the bid due date when it makes sense
4. Transmit extra copies of the RFP
5. Conduct site visits and prebid meetings

Even though a clear and concise RFP is essential, it is virtually impossible for all aspects of a proposed contract to be reduced to written form in a manner that precludes error, misunderstanding, or uncertainty. Other communication efforts are needed. Don't rely on the documents, and don't assume that people read them as carefully as you do.

CONTROLLING COMMUNICATION RISKS

A common problem is uncontrolled, inconsistent, or undocumented communication from bidders to owner and from the owner to bidders. This results from many factors; chief among them is the owner's failure to designate one person as responsible for answering or responding to bidder questions or requests for information. The contract manager is an ideal candidate for this duty. Although he or she may not be qualified to answer specific questions, such as those dealing with design specifications, material substitutions, construction scheduling, and the like, the manager can receive all incoming communications, get an appropriate reponse, and transmit it to all bidders equally. All questions and other communications can then be formalized through the addendum process. This keeps all bidders on equal footing, and information advantageous to each is available to all.

Common sense dictates that a single point of communication be established and that the bidders be told that all questions or other correspondence be funneled through that source. Only through this single-point philosophy can control over the dynamic bidding process can be maintained. More often than not this is abused, with bidders communicating directly with designers, consultants, owner management, and others and the result being misunderstanding and confusion. The contract manager either corrects them or suffers the consequences. Better to do all you can to prevent them.

ISSUING ADDENDA

An addendum should be sent to all bidders when changes are made to the content or intent of the RFP. Significant clarifications should also be recorded and transmitted to bidders through addenda. In order to provide control over the addendum process, each addendum should be dated, numbered, and explicit in its meaning. An example addendum letter is shown as Exhibit 7. As shown in the sample proposal form (Exhibit 3), acknowledgment of addenda, by number and date, should be indicated with the submitted bids.

Because addenda are often misused or misinterpreted, control over their use cannot be overemphasized. The following suggestions should help avoid commonly encountered problems.

1. Use only one type of document to modify the RFP, and modify the RFP *only* through that document. Avoid multiple change instruments. When modifications or exceptions are made verbally, through letters to bidders, or otherwise, make sure they are included in an addendum issued soon thereafter.

2. Incorporate changes made by addenda into the subsequent contract documents once a successful bidder is chosen. Do not incorporate all addenda issued during the bid cycle "by reference." This makes interpretation and administration of the eventual contract difficult. Often a plethora of addenda are issued, and it is not uncommon for one addendum to modify, enhance, or extend the changes made by previous addenda. To ask those responsible for contract administration not only to understand the actual contract documents but also to wade

EXHIBIT 7. Sample Addendum

ADDENDUM

(Date)

(Bidder's name and address)

(Owner's name, project, contract description, and contract number)

ADDENDUM NO.
(addenda are numbered consecutively for each RFP, beginning with number 1)

Please refer to our letter dated (date of invitation to bid letter that transmitted the request for proposal to all bidders) requesting your bid for (contract description and contract number) for the (project name).

Our request is hereby amended as follows: (for second and following addenda, use "Our request, as amended by Addendum No. _____, is hereby further amended as follows:")

(Revisions should be made by document type, as the following examples illustrate.)

A. The following revisions are made to the Proposal:*

1. Page 6 of 8, under "The Bidder requires the following temporary facilities," revise Item 1 to read "Construction Water Requirements."
2. Page 7 of 8, add as fourth paragraph: "A complete résumé of the proposed superintendent shall be submitted with Bidder's Proposal."

* Changes to the schedule of prices should be made by issuing replacement sheets for each page to be modified. These sheets should be identified by "Addendum No. _____" under page numbering. When several pages of the proposal are revised, it is recommended that an entirely new proposal section be issued.

B. The following revision is made to the Special Conditions:
1. Page 3 of 17, Article 4.1.6, add the following:
    "8. Furnishing and installing level indicators"

The bid due date of _____ shall remain unchanged (or) is hereby changed to _____ .

Except as specifically revised by the terms of this Addendum No. 1 (or) by

EXHIBIT 7. *(continued)*

Addenda Nos. _____, _____, all provisions of the Request for Proposal as originally issued shall remain in full force and effect.

Kindly acknowledge receipt of this Addendum and that you have taken all provisions hereof into consideration in preparation of your bid by signing in the space provided below and returning this Addendum with your Proposal.

Very Truly Yours,

_____
(owner's signature)

Received by:

_____   _____
(name and title of bidder)   (date)

---

through all issued addenda invites misinterpretation and serious performance risk. Since the conformance of the contract represents the owner's last chance to start with a fresh set of documents, make sure this is done. Subsequent "contract changes" will make interpretation difficult enough.

3. Attempt to make addenda brief, concise, and easily understood.

4. Issue addenda in a timely manner. Let bidders know of changes as soon as possible so that the addenda's full impact may be understood and reflected in the bids.

5. Send addenda only to those who originally received the RFP. Do not get involved with subtier information responsibilities. Addenda should state that it is the bidder's responsibility to transmit changed requirements to other affected parties (subbidders, material vendors, and so on). Make sure that addenda are circulated internally within the owner's organization, particularly to those preparing fair price estimates, making schedules, ordering material, and so on.

6. Always reference the affected RFP, by section, page, paragraph, and/or line number. If a requirement is given in more than one place, make sure it is changed in each location. This is another reason to avoid repetitive or redundant requirements in the RFP.

7. Often, significant questions will cause the owner to clarify but not change existing RFP material. When this is done, clarifications or enhanced explanations should be segregated from actual changes and identified as such.

8. Always refer to the bid due date, even if it is unchanged by the addendum in question. This is the first question bidders will have when receiving addenda: is the bid period extended?

9. Number addenda consecutively and always refer to previously issued addenda by number and date. This helps all parties keep track of previous changes. If a bidder failed to receive an addendum, this should be discovered as soon as possible.

10. If massive revisions to a section are needed, consider issuing replacement sheets or a new section incorporating thoses changes. To ensure that the new sections or sheets are used in the final bid, number or otherwise identify those sheets before sending them to bidders. For example, print them on different-colored paper.

11. Require receipt acknowledgment. Some owners send two copies of the addenda, asking the bidders to return a signed copy upon receipt. Others ask bidders to list the addenda in their proposal. Both are recommended.

12. If addendum changes are significant, contact the bidders immediately by phone or fax. Let them know the changes are coming as soon as possible, but always follow this notification with the addendum itself.

REQUESTS FOR ADDITIONAL DOCUMENTS

Sometimes owners are requested to provide additional copies of the RFP documents, particularly the drawings. A good practice is to respond only to bidder requests—to inform subbidders to contact the prime bidder in question for additional copies. Most owners establish policies dealing with the cost of these documents, either absorbing it themselves or charging the requesting party. As with requests for additional documents, subbidder questions regarding the RFP should be directed through bidders. Once again, this is done to reduce annoying or frivolous inquiries and to maintain a lack of association with any groups other than the invited bidders. Inconsiderate or lazy bidders can easily take advantage of your courtesy.

EXTENDING THE BID PERIOD

A decision to extend the bid period should be based on significant occurrences that justify the postponement, rather than on the convenience of a bidder. If the RFP is well prepared, a sufficient and appropriate bidding period has been scheduled, and no extenuating circumstances arise, the bid date should be held firm, despite the occasional protestation of an overworked bidder. Many contractors will request an extension as a matter of course, hoping to ease their estimating workload or to give priority to other bids for other projects. Of course, the length of the original bid period should take into consideration the complexity of the scope of work, firmness of technical information, number and complexity of pricing alternatives, and other factors affecting the time required to prepare and submit responsive bids. Bid periods would naturally vary from contract to contract.

Some factors that may require an extension are:

1. Discovery of significant errors or omissions in the RFP.
2. Late release of the RFP to bidders or failure of any bidder to receive the RFP in sufficient time to respond.
3. Majority of bidders claim not to be able to meet the due date. It makes no sense to stand firm on the original due date if no bids will be received by then.
4. Changes in the project schedule that cause delay of the construction start-work date (bids are not needed as early as originally planned).

When these or other causes require an extension, all bidders should be notified by telephone or fax followed by a written addendum. If some bids have already been received, those bidders should be allowed to retrieve and resubmit their bids at the later date. This is another reason to refrain from opening bids as soon as they are received (instead, open them simultaneously on the bid due date).

QUESTIONS OF PROPRIETY

Because the bid cycle represents a period when the owner is probably most susceptible to impropriety or the accusations of impropriety, specific precautions are in order. These include owner policy on contact by its employees or representatives with prospective bidders, the acceptance of gifts or gratuities, and other activity that may impugn the objective receipt and evaluation of competitive bids. Often bidders will ask to call on the owner to discuss the RFP, review drawings, or simply chat during the bid cycle. This and all other actions that affect the perception of propriety are not recommended.

The owner and its agents should maintain an arm's-length approach toward all prospective bidders and contractors. This advice pertains to the administration period as well, for the evaluation of claims or change order quotations may be impugned by familiarity beyond that which is absolutely required to conduct the business of contracting.

PREBID CONFERENCES AND SITE VISITS

So that all bidders are familiar with the intended scope of work, project requirements, and rules of the bidding process, it is usually advisable to conduct prebid conferences and require all bidders to visit the site and acquaint themselves with conditions that may affect their performance, if selected. Site visits are handled in one of two ways:

1. Allow bidders individually to schedule their visits with the owner's site manager and conduct independent site visits or tours.
2. Require all bidders to visit and tour the site at a common, specific time.

Oftentimes, when the project site is under development and facilities exist, the site visit is combined with the prebid meeting. The meeting takes place at the site and is followed by an owner-conducted tour of the project, sometimes called a "job-walk." The site visit is to acquaint the bidders with such things as access, storage and laydown areas, spatial restraints (crowding), and weather. It also counters possible claims at a later date by contractors who feel they were not informed or aware of performance-affecting factors (unknown site conditions).

The prebid conference, on the other hand, is a good-faith effort by the owner to inform, clarify, or explain RFP content to all the bidders at the same time rather than going through a cumbersome process over the telephone or through the mails. All bidders are invited, with their subbidders. An agenda is sometimes used, and bidders are encouraged to ask questions or request clarification of the RFP at the meeting. To be meaningful, it should be scheduled midway through the bid cycle, or at least after the bidders have received and had sufficient time to become familiar with the RFP. If conducted properly, the prebid conference is a valuable control technique for contract formation. Problems will occur, however, if it is not managed well. Here are some recommendations:

1. Conduct a joint site visit. Do not schedule numerous visits only at bidder convenience. This assures consistency of information and is easier to manage.
2. Prepare a strict prebid conference agenda and stick to it. Avoid extraneous, off-the-cuff or sidebar remarks that may confuse bidders.
3. Rehearse the conference. Make sure each speaker addresses only those topics assigned to him or her; if a question is beyond his expertise, defer to one who is responsible for that aspect. The construction manager should not answer design questions, and the design engineer should not answer contract administration questions.
4. Do not allow changes to the RFP *except* through written addenda. Inform the bidders that any changes resulting from the conference will be implemented through written addenda following the conference and that these will constitute the only changes.
5. Keep control of the meeting. Do not allow simultaneous or parallel discussions. Discourage side conversations before or after the meeting.
6. Keep minutes for your own purposes but do not distribute them to bidders or incorporate them into the RFP by reference. Use the addendum process.

7. Do not allow bidders to "dry run" their proposals. Bidders may test your acceptance of offerings at the meeting.

8. Present a united front. Argue in advance as to who speaks for the owner. This will minimize having to correct one another in front of the bidders.

9. Avoid the temptation to answer all questions on the spot. If a question cannot be answered immediately, say so. Clarify it later in an addendum.

10. Distinguish between actual changes to the RFP (and therefore candidates for addenda) and mere questions of interpretation.

11. Review and summarize the meeting, areas of change, expected addenda, and so on, prior to adjournment.

12. Maintain a professional attitude at all times. Do not show favoritism toward any bidders. Avoid casual conversation and the appearance of impropriety; that is, do not go to lunch with a bidder, drive to the meeting with one, or have drinks afterward with one.

REPRESENTATIVE CASES

1. During a crowded and boisterous prebid conference, a question regarding sales tax is addressed to the speaker, the owner's mechanical superintendent. The superintendent, thinking of a previous project on which sales tax was paid by the owner, answers that sales tax should be excluded from the bid amount, and the discussion quickly turns to another topic before adjourning. The contract manager, aware that on this project sales tax is to be paid by contractors, spends the remainder of the week trying to contact each bidder and correct the misstatement.

2. As the contract manager is entering her car in the parking lot after a prebid conference, an attendee, who identifies himself as a representative of bidder C, stops her and states that due to the apparent confusion on the project, his company does not want to bid. The contract manager therefore sends no further addenda or otherwise communicates with bidder C. As it turns out, the prebid attendee, formerly of bidder C, had just negotiated an employment agreement with bidder B, a former competitor, at the prebid conference, and bidder C is vitally interested in the proposed work. Insist on written regrets.

CHAPTER 9

BID RECEIPT AND EVALUATION

When completed bids are received, two interrelated objective should be met:

1. An impartial, objective evaluation of all bids, including such aspects as price, technical content, and commercial considerations, should be made so that the best available offering is chosen for award.
2. The security of confidential bid and bid evaluation information should be maintained at all times.

To achieve these objectives, the bid receipt and evaluation process needs to be well planned in advance. The evaluation process should be structured, with owner personnel and their representatives made fully aware of their roles. For major, complex scopes of work some owners appoint a bid evaluation committee to perform the review and make an award recommendation. But for most situations management, representing various disciplines—cost engineering, contract individuals, design engineering, and so on—perform separate evaluations—with the contract manager acting as coordinator.

BID EVALUATION PROCESS

A general bid evaluation process, containing separate commercial, technical, and cost reviews leading to a consensus of opinion and eventual award, is represented in Figure 14. When the RFP has been carefully prepared and the bid process managed well, the evaluation is usually a straightforward process. But it can be complicated by (1) ambiguous RFP and specimen documents, (2) an unmanaged bidding cycle, (3) a large number of bid alternates or complex pricing arrangements, and (4) an uncontrolled evaluation

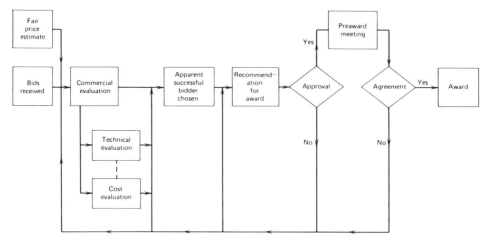

FIGURE 14. Bid evaluation process.

methodology. Perhaps the most common difficulty is a lack of definite responsibility assignment—where the respective roles of owner or agent personnel are not clarified prior to bid receipt.

Most owners document the results of the bid evaluation on some sort of "spreadsheets" that list the bidders, detail the contents of each bid, and record the data leading to an award recommendation. This is a good practice, for it is not unusual for the process to be repeated more than once for the same contract (should an agreement not be forthcoming with the apparent successful bidder). It may be necessary in these cases to return to the other bids to make a second choice. Well-documented evaluations allow this to proceed rapidly, and speed is often of the essence—valuable time has been used for the original evaluation process.

For unit-price and lump-sum bidding, the price evaluation is straightforward. Estimated units (quantities) can be used to extend unit prices to anticipated costs, and a "bottom-line" cost to the owner can be determined. Base bids should be evaluated along with optional or alternative bids. Nonprice information should also be documented on such a form. This may include the identity of proposed subcontractors, schedule durations, bonding company, or performance data of proposed equipment.

The contract manager usually employs the aid of a technical specialist (such as the engineer responsible for the technical specifications and drawings) to review the bids and accept or reject each depending on technical criteria without regard to (or considering) price. Ideally, only those needing price data should have access to it. However, some work requires that cost and technical information be studied interchangeably—such as to determine cost trade-offs between alternative systems, components, or equipment. But to the extent practical, cost-sensitive information should have limited visibility.

In many cases first costs are not the only costs the owner incurs. Additional costs that may affect the overall decision include "evaluated costs" or "imputed costs" to cover such areas as:

1. Evaluated costs
   a. Equipment performance penalties or credits
   b. Cash timing and terms of payment
   c. Escalation provisions
   d. Optional pricing alternatives
2. Life-cycle costs
   a. Energy requirements
   b. Annual operating costs and maintenance costs
   c. Replacment intervals
   d. Expected salvage value
   e. Reliability (expected costs of failure)
3. Nontechnical costs of performance
   a. Insurance and bonding (if these are to be for the owner's account)
   b. Retention terms
   c. Schedule incentives and penalties

If any of these cost considerations are to be made, the owner should identify them in advance and appoint people responsible for determining their impact, and have objective evaluation criteria established in advance of bid receipt. An extensive list of bid evaluation considerations is shown in Exhibit 8.

BID SECURITY

When bidding is by invitation, the owner should see that pricing and other confidential information contained in those bids are secure. Other bidders should at no time have access to data contained in competitors' proposals, nor should any bidder be apprised of the evaluation status prior to award of contract. Failure to secure confidential information has several detrimental effects:

1. It can reduce the number of interested bidders for future work.
2. It can cause ill feelings among the owner, bidders, and eventual contractors.
3. It can result in recriminations, questions, and harassment from unsuccessful bidders who feel they should have received the award.
4. It can weaken subsequent negotiation efforts with the apparent successful bidder.
5. It can affect the impartiality of the evaluation process itself.
6. It can lead to public scandal, excessive media attention, and everything these bring.

Exhibit 8. Bid Evaluation Considerations

FIRST COSTS

1. Lump-sum prices
2. Unit prices extended by estimated amounts
3. Unit prices for adds and deducts
4. Percent markup for cost-plus extra work
5. Cost of bonds, premiums for insurance if paid by owner
6. Escalation

COMMERCIAL CONSIDERATIONS

1. Warranty, guarantee
2. Payment terms
3. Retention terms
4. Cancellation charges
5. Discounts
6. Service contract fees
7. Location, size of service force
8. Cost of technical representatives and/or training during construction, testing, initial operation
9. Schedule incentives, penalties
10. Cost penalties, incentives
11. Contractor's performance history
12. Pricing alternatives
13. Proposed subcontractors
14. Proposed schedule
15. Present workload
16. Proposed staffing and resources available

INTANGIBLES

1. Contractor reputation
2. Responsiveness to RFP (exceptions taken, etc.)
3. Aesthetics (appearance, style, size, color, etc.)
4. Stability of contractor, subs, vendors
5. Community representation
6. Social/political impacts
7. Environmental impacts not quantifiable but actual
8. Collateral relationships with contractor, subs

LIFE-CYCLE COSTS

1. Annual power, fuel requirements
2. Estimated replacement intervals
3. Replacement costs
4. Salvage value
5. Maintenance costs
6. Labor requirements for operation
7. Spare parts needed
8. Maintenance tools, equipment needed
9. Reliability: downtime, frequency, cost of alternatives while down

TECHNICAL CONSIDERATIONS

1. Performance incentives, penalties
2. Material used
3. Component parts, systems, vendors
4. Space requirements
5. Performance capabilities
6. Auxiliaries required
7. Items eliminated or made less expensive
8. Operating training required
9. Compatibility with other equipment, structures, systems
10. Upgradability
11. Technological lifespan
12. Ease of operation
13. Licensability
14. Sound, vibration, pollution, and other environmental parameters

EXHIBIT 8. *(continued)*

15. Safety of construction and operation	18. Requirements for construction: power, water, space, impact on collateral work
16. Configuration, accessibility	
17. Availability of spare parts, tools	19. Environmental impact

To avoid these problems, issue strict bid security procedures and enforce them at all times. When others outside the owner's organization are involved in the evaluation—such as A-E firms and consultants—the difficulty of ensuring bid security increases. The following suggestions should help protect the confidentiality of bid information during the evaluation process and afterward.

1. Bids should remain sealed until all are opened on the due date. Resist the temptation to get a head start on the evaluation effort and to open bids as they are received.
2. Bids should be opened together in the presence of only selected individuals. Some owners insist on the presence of a disinterested individual as witness to the bid opening. This person also signs a log attesting to the content (and amount) of bids received. This is done as a check against improper handling by those responsible for opening bids.
3. While being evaluated, opened bids should be attended at all times.
4. When not in use, bids and evaluation material should be locked in a secure location. Do not forget computer files and databases.
5. Evaluation information should not be disclosed to others (internal as well as external) before an award made.
6. When possible, nonprice reviews should be performed with unpriced bids (such as for technical evaluations of proposed equipment).
7. Once the evaluation is complete, bids and evaluation material should be locked in inactive files for a nominal period of time. They should not be destroyed until long after the contracted work has been completed. Transfer computer files to disks and lock them with the paperwork. Erase resident files.

USE OF A FAIR PRICE ESTIMATE

A fair price estimate, or internally prepared bid estimate, is a valuable control and checking technique used during bid evaluations. It works like this: The contract manager requests the cost engineer (or estimator) to "bid" the work using the identical RFP documents sent to actual bidders. The estimator is given the same period of time to prepare the estimate as bidders are given to bid, with no other information or insight than is provided bidders. The resulting fair price estimate is prepared on the identical pricing forms and to the identical format as actual bids, allowing it to be used easily as a benchmark against which to measure actual bids. Significant benefits of a fair price estimate include:

1. It helps to surface bidder errors or misunderstandings. A big difference between the fair price estimate and actual bids may indicate that the RFP and/or prebid meeting did not clarify the scope of work or pricing conditions. Should all bidders have the same misunderstanding, this may not otherwise be detected.
2. It helps the owner isolate variances or exceptions among the actual bids.
3. It points out subjects for preaward negotiation.
4. It identifies unbalancing of bids (front-end loading, addition and deletion unit pricing, etc.).
5. It is a guard against bidder collusion or bid rigging. When actual bids are very close to one another in price, the appearance of healthy competition may be deceptive. The closeness could represent a deliberate attempt on the part of all bidders to "fix" the price.

For a fair price estimate to bring these advantages, it should be prepared "blind." That is, the preparer should not see the genuine bids before preparing his or her own, and no inside information should be provided.

SELECTION AND NOTIFICATION

On rare occasions, a clear bid winner is determined, no exceptions or qualifications are contained in its bid, it is technically qualified, and an award can be easily made without postbid discussions or negotiations. It usually does not happen this way, and even though an *apparent* successful bidder is determined, specific issues must be resolved before an award is advisable. Preaward meetings are held to resolve these issues.

When this happens, the bidder should not be told that it is the low bidder, or even the apparent successful bidder (most will assume this in any case). Some owners refer to these meetings as "postbid" rather than as "preaward" for this reason. The semantics do not hide the intention—bidders know what's up. If the discussions or negotiations are successful, award can then be made. Only at that time (and not before) should unsuccessful bidders be notified, in letter form, that they were not chosen. The owner may or may not wish to let others know who was chosen or why at this time, but this will become apparent to all shortly. The specifics of the evaluation—who was second, third, fourth, by how much, and so on—need not be disclosed.

ERRORS AND REBID

Errors in bids are common, and sophisticated owners realize that to accept a bid with admitted errors—to force the bidder to perform—would jeopardize the project and may not even be legal or possible. If errors substantially affect the price and order of bids, a rebid should be considered. The bidder in error should be offered a chance to correct its errors, and the bids should then be reevaluated, if required. This is another reason to hold off opening any bids until all have been received. It is much easier to allow bid retrieval (such as when a bidder suddenly discovers it made a price mistake) before bids have been opened than afterward.

If significant RFP errors surface after receipt of bids, or if the proposed scope of work is seriously altered, it may prove wiser to rebid the contract— send out new RFPs to all bidders and repeat the bid process—than to award a contract to the apparent successful bidder and attempt to negotiate the subsequent changes.

REPRESENTATIVE CASES

1. The chief purchasing agent for a large electric utility had the habit of opening contract bids as soon as they arrived in order to get a headstart on bid evaluations. For a large mechanical contract, two of four bids arrived in the mail 10 days prior to the due date, and the purchasing agent had his staff begin preparing spread sheets for the evaluation. Lump-sum prices for the two early bids were:

Bidder A: $7,298,660

Bidder B: $8,925,000

Two days prior to the bid due date, the construction manager called for a one-week bid extension. Since no changes were made to the scope of work, bidders A and B allowed their bids to stand. On the revised due date, the remaining two bids arrived. They were:

Bidder C: $10,100,500

Bidder D: $ 7,298,600

The owner awarded the contract to bidder D, who was suspiciously only $60 under bidder A. Bidder A complained bitterly to the utility and the press that the security of its bid was compromised. The resulting furor delayed contract award and start of work for four months.

2. A cost engineer for an A-E firm handling procurement for the project of a long-standing client left his desk to attend a meeting on bid security. He also left the bid evaluation sheets he was working on opened across his desk top. A competing bidder, visiting the office to distribute brochures to a buyer across the hall, happened to notice that her company's bid was low by $97,000. When she returned to her own office, the contractor's representative

had her estimator call the A-E firm and report a $90,000 mistake in her firm's bid estimate.

3. The vice president of operations for a major manufacturing firm was irate when he called the project manager of its consulting engineering firm. It seems he had just received a call from a project bidder who knew the exact amount of each of the seven bids now being evaluated by the engineering firm. During the next few days, the leak went undetected, but the client stressed that it must have come from the consultant's shop, as the client, though privy to the bids, adhered to a strict bid security procedure. Though aware of the need for security, the engineering firm had no formal procedures in that regard. As a result, the client removed procurement services from the engineering firm's scope of work—at a substantial reduction in its fees for the project.

4. A major mining and minerals company would evaluate lump-sum construction bids by choosing the low bidder and having a technical evaluation made of that bid alone. On a proposal for coal handling equipment, bidder A was awarded a furnish-and-install contract at $2 million. After five years of operation, the mining company's accountant determined the actual cost to be $5 million as follows:

Purchase price:	$2,500,000
Fuel costs:	200,000
Maintenance:	700,000
Spare parts:	400,000
Downtime costs:	1,200,000
Five-year cost:	$5,000,000

On checking further, the accountant found that another mine of the same company had bought the same type of equipment at about the same time. The second mine, however, had chosen equipment from bidder B for $3 million. The cost experience on this equipment was:

Purchase price:	$3,000,000
Fuel costs:	100,000
Maintenance:	400,000
Spare parts:	300,000
Downtime costs:	400,000
Five-year cost:	$4,200,000

5. For an eight-mile oil pipeline project an RFP was released for furnishing and installing buried piping on a lump-sum basis. Bids were:

Bidder A:	$4,100,000
Bidder B:	$4,050,000
Bidder C:	$4,150,000

Bidder D: $4,000,000
Bidder E: $3,950,000

The construction manager, satisfied that a competitive and representative set of bids had been received, scoffed at the need for a fair price estimate. Since one had been prepared, he read its amount of $3.1 million in amazement and ordered his cost engineer back to her computer for a recheck. The result was the same. A preaward meeting was called, and bidder E was asked to provide its estimate units of pipe, connections, cut and fill, and so on. When the cost engineer extended these numbers by the unit prices furnished in E's bid for adds and deducts at the meeting, she arrived at the price of $3 million. Bidder E, visibly flushed, called for an adjournment to telephone his office. When he returned, bidder E explained that a "clerical mistake" must have occurred. His revised bid, he allowed, was $3.1 million. He was awarded the contract for that amount and performed remarkably well. Although never proven, an official of bidder E later intimated to the contract manager that all five original bidders held their own "prebid" meeting prior to submitting bids (no doubt at the "No-Tell Motel"). They decided to add $800,000 ot each of their bids. Without the owner's fair price estimate, the bids would have appeared competitive, and the bidders' collusion would have gone undetected—and cost the owner $850,000.

6. To qualify for exclusion from clean air standards for a coal-fired electric generating plant under a grandfather clause, foundations for both 400-megawatt units needed to be solicited quickly. Contractor A emerged as apparently successful and was called to arrange a preaward meeting in April—eight months before the deadline for completion of the affected foundations. Since construction of the foundations would require seven months, contractor A, knowing the extreme urgency on the part of the owner, procrastinated. It continually found reasons for postponement of the meeting, and therefore delay of contract commencement. As time passed, the criticality of schedule considerations grew, until contractor A finally arrived at the negotiation table with time on its side. The project owner was under such pressure to begin work (a $200 million potential savings without the need for certain pollution control equipment compared with a $3 million excavation and foundation contract) that it met virtually every demand the contractor presented. Contractor A was awarded the most favorable contract in its history.

CHAPTER 10

CONTRACT AWARD

Once a successful bidder has been selected and both parties sign the contract documents, an award has been made. Provided that no unresolved issues remain and no significant changes have occurred during or after bid evaluation, the award process is usually a simple exercise. Contract documents are prepared to reflect the specimen terms contained in the RFP as well as the contents of the proposal submitted by the apparent successful bidder, along with any last-minute changes.

The formal execution of the contract documents, then, signifies the end of contract formation and the beginning of the next major phase of contract management: contract administration. As simple as this exercise may seem, there are usually many tasks required to ensure that it is performed in a timely manner. And there are many pitfalls to recognize and avoid during this critical period. Many of these stem from the time pressure involved—the need to mobilize and begin construction work without further delay. Owners who do not allow sufficient time for the formation period, who use poor contract and RFP documents, who do not control the bid cycle often find themselves in compromising positions during contract award. They accept imprudent terms and conditions, compromise on control and quality standards, or, worst of all, order the contractor to start work without the protection of signed documents.

THE ROLE OF NEGOTIATION

The term *negotiation* conjures up visions of a group of opposing parties, sitting across a table from one another and engaging in lengthy, high-stakes bargaining sessions. Many of those who have not been involved in these exercises view negotiation as a glamorous adventure, full of risks to be taken, points to be made, or sizable gains to be won or lost depending on one's "negotiation skill." In a way, this view is correct—negotiation is a highly subjective and often unpredictable gamble. As such, it pays to enhance one's negotiating skills, but more important, to avoid the need for negotiation entirely.

Clear, concise owner requirements (properly designed and interpreted RFP and specimen contract documents), an adequate number of qualified bidders, and a disciplined bidding process are major factors reducing the need for preaward negotiation. The more nebulous the documents, untenable the owner's position during formation, or unreasonable the owner's price expectations, the greater the danger of relying on negotiation as a formation tool.

In general, contractors who have been in the business for any length of time are much more adept at negotiation than most owners they will encounter. They engage in contract negotiations continually—with subcontractors and suppliers as well as other owners. Most owners face negotiation of major contracts infrequently, are not sufficiently trained or skilled to achieve success, and often fail to do so. They have little to gain and much to lose. Wise owners realize these drawbacks and avoid the need for negotiation whenever possible. Besides the risk involved, negotiation seldom results in an amicable agreement, with both parties pleased with the outcome and anxious to begin their contractual obligations on good terms. Whether immediately or soon thereafter, one or both parties may feel cheated, outmaneuvered, or in a poor bargaining position for future negotiations, which are bound to take place during the course of contract performance. This is not the recommended frame of mind at the beginning of a long and challenging project.

When negotiation is unavoidable, owners should prepare themselves well. Negotiation team members should be chosen and their roles and responsibilities fully understood. They sould be vested with the authority to conduct the negotiations—that is, to come to agreements, make decisions, and commit the owner when necessary. The same should be required of the contractor. Nothing is more frustrating than reaching a long-attempted agreement with the opposite party only to find that it must obtain the authority to fulfill the commitment at some higher level.

Negotiable elements of the proposed contract should be distinguished from nonnegotiable ones before the discussions begin. Owner preparation should include rehearsals, creation of a strict agenda, and discussions and/or approval of possible agreements with higher owner management, if needed in advance. Careful records should be made of the negotiation results, and these should be formally incorporated into the subsequent contract documents. Every preaward understanding should be reduced to written form and made part of the contract before the award is made.

One final suggestion regarding negotiation involves the use of bidder or contractor feedback to contract documents or formation procedures. Should an owner find itself continually negotiating on the same terms or conditions, or with the same set of bidder objections or exceptions taken to RFP documents, time may have come for a review of these terms and conditions. Some may be unreasonable, unclear, or even unnecessary. If so, toss them out or revise them to reflect current practices in the industry. To ignore these signals—that the contract documents are, in effect, lacking—is to risk the increased time and cost required continually to clarify and negotiate from an untenable position. Quite often the terms in question do not actually offer owner protection or are not enforceable anyway. Contractors, knowing this, may accept these terms only as a compromise for the owner's acceptance of very real, meaningful, and enforceable terms favorable to the contractor. Don't fight for something you do not need or will not get anyway—you may give up some things you need or are entitled to in the "bargain."

BID SHOPPING

Bid shopping refers to the practice of "auctioning" the contract award to the low bidder. When it is used, the owner contacts one bidder and confronts it with the lowest bid. It then asks the bidder present if "it can beat this price," with contract award hinging upon the bidder's response. If the answer is no, the owner contacts another bidder and asks the same question, until the "counteroffer" by the owner is accepted. And it is not uncommon for the owner, once a contractor has agreed to beat the stated price, to call the other bidders for another round of this exercise, asking them if they can beat the new price. It is a ratcheting process, with the price being gradually squeezed down to the lowest level acceptable to one bidder.

Most sophisticated owners do not practice bid shopping. Any minor price reduction achieved through these means is greatly offset by the serious disadvantages it presents. Among these is a contract agreement that begins with bad faith. Contractor performance is bound to suffer when award is made in this manner. Quality is often shortchanged, and subsequent agreements for such things as change orders, schedule acceleration, and the like are difficult when the contractor feels that it has "left money on the table" prior to award. Other disadvantages arise beyond the immediate contract. Owners who gain a reputation as bid shoppers will find that the ensuing bids are padded by most contractors in anticipation of the price reduction they expect through bid shopping. And many contractors may simply refuse to bid on contracts when they know bid shopping will occur. This reduces the number of bidders and therefore reduces healthy price competition.

Bid shopping is often used by contractors when obtaining quotations from suppliers or subcontractors. But owners who expect qualified contractors to bid in good faith should avoid this practice. A well-prepared RFP, adequate number of qualified bidders, and thorough evaluation process should achieve more acceptable results.

CONFORMING THE CONTRACT

Using the original RFP documents, minus those that are for bidding purposes only—the invitation and proposal forms—the contract manager prepares the actual contract documents by:

1. Incorporating all changes made through addenda during the bid cycle
2. Transferring bidder submittals and other variable data to the agreement form (prices, schedules, methods, procedures, equipment data, and so on)
3. Incorporating changes and understandings made during preaward and negotiation sessions

The resulting contract documents, together with the drawings, constitute the conformed contract. Several copies are made and sent to the contractor for signing (execution). Once returned, they are signed by the owner and distributed to the contractor and others within the owner's organization. It is recommended that the contractor sign before the owner unless both sign at the same session, for some contractors have been known to alter the documents before signing them and returning them to the owner. Detailed scrutiny of each page of the documents can be avoided by following the suggested sequence. This does not happen very often, but why take the chance?

Often there are ancillary documents requiring signatures. These may include bond forms, project labor agreements, or waivers of lien. These should be reviewed for proper completion and signature before the owner signs. As mentioned earlier, incorporation of the bidder's proposal by reference (rather than transferring necessary data to the owner's agreement form) is not recommended.

Copies of the executed documents should be filed for record in the owner's office and distributed to those having a need within the owner's corporate or project organization, particularly field construction management and contract administration personnel. Since the bid drawings are now part of the contract documents as well, some owners require both parties to initial and date one or more copies of these, indicating they are the agreed drawings of record for the contract. Because a flurry of drawing revisions often follows contract award, the practice of establishing a "base set" of drawings is good. Contractual changes required by drawing revisions can be easily determined when they occur.

LETTERS OF INTENT

Although conformance of the contract documents should be quick and easy (given proper contract formation controls), some owners cannot produce an acceptable set as quickly as they wish to have the contractor mobilize or begin working at the site. In these cases, resolution of any minor negotiations or the processing of paperwork is not allowed to impede contract award. An interim contract, called a letter of intent or letter contract, is issued the contractor. It references the contract, the proposal, or RFP and orders the contractor to proceed with the work pending issuance of the actual contract documents.

Even when they are used sparingly, under carefully controlled conditions, and are precisely worded, letters of intent have limited value. Otherwise, they are extremely dangerous and should be avoided, if not prohibited. They are frequently relied upon as an excuse to delay the award of the contract or resolution of troublesome differences between the parties—differences that must be addressed before a contract can be signed. Indeed, some would argue that the mere possibility of a letter of intent actually encourages procrastination in resolving problems—the letter of intent relieves the immediate need for their resolution. Letters of intent should not be used as a matter of course or abused by allowing reconciliation prior to a contract to be delayed.

The major problem with letters of intent is that, for sake of brevity, they reduce the contract documents from many pages (usually several hundred) to one or two. This cannot be achieved without loss of owner (and contractor) protection. A contract thus formed exposes the owner to severe commercial and technical risks. If letters of intent are used because of a paperwork backlog, this problem should be corrected. When they are used because differences are not yet resolved, these differences should be faced. Perhaps they are incapable of resolution, and a contract between the two parties is unachievable. If so, embarking on an unachievable contract with scant protection is the greatest of errors.

For those owners who insist on using letters of intent under these circumstances, the letters should be carefully drafted. They should be numbered, dated, incorporate the accepted terms and conditions of the RFP by reference, and be countersigned by the contractor. If any bonds are required, they should also be received. Insurance certificates should be submitted by the contractor before commencing work. Scope and price limitations are also strongly recommended and should be specified in the letter. All letters of intent should state that they are temporary in nature and be effective for only a certain period of time, and include an expiration date 30 or 60 days later. If the letter of intent is about to expire and a contract has still not been drawn and signed, an extension of the letter can be issued—in effect for another stated period of time. This achieves two important objectives: it limits the owner's cost exposure, and it forces owner management to periodically review conditions retarding proper formation. The pending expiration serves as a "tickler" to those responsible for resolving the issues in question and executing more adequate contractual documents.

Despite all the risks associated with letters of intent, their use persists. It is not unheard of for an entire contract scope of work, spanning one or more years to complete, to be performed under a letter of intent—that is, an executed contract is never achieved. The goal of those responsible for contract formation should be to schedule and control their activities so as to eliminate the need for and the exposure caused by letters of intent. These instruments probably represent the single most dangerous practice associated with the entire subject of contract formation.

PART 3

CONTRACT ADMINISTRATION

As opposed to the activities of contract formation, where a fairly linear sequence of events takes place for each contract to be awarded, contract administration represents a collection of processes, responses to events, and controls that are not as easily arrayed or performed. Some requirements of contract administration are common for all contracts, such as progress payments, record keeping, and contract closeout. However, there may or may not be a need for other business transactions and controls. Claims, change orders, backcharges, and short-form contracts represent some topics in this category.

This part presents the broad topic of contract administration by discussing both sets of circumstances: those that occur for all contracts, and those that typically (some would say unfortunately) occur during the contract performance period of many contracts. Major topics include:

Mobilization and Commencement. Initial meetings and communications between the contracted parties are described, along with a system for maintaining critical contract documentation, correspondence, and submittals.

Progress Billings and Payments. Periodic progress payments are a way of life in the construction industry. We discuss ways for them to be determined objectively, in compliance with the intent of contract documents, retain owner leverage over the work, and processed in an efficient and timely manner.

Change Orders. Nothing is more constant than change in construction work. This chapter describes typical causes of change and ways an effective change control program is implemented.

Backcharges. This chapter describes typical conditions provoking the need for backcharges, a procedure for cost recovery, and now backcharges can reduce total project cost.

Claims. Claims are discussed from the owner's perspective, whether it is (1) defending against a construction claim brought by a contractor, or (2) pursuing a claim against another party. Claim causes are analyzed, and typical components of contractor claims are addressed.

Short-Form Contracting. Some owners allow a certain degree of flexibility needed to respond to unforeseen events through the use of short-form contracts with certain scope or dollar limitations. This chapter describes their use and abuse.

Contract Closeout. This chapter addresses the wrapping up of unfinished business at contract termination. Final payment, delivery of remaining products and information, and postperformance controls are presented.

Contract administration begins with the signing of contractual documents, continues throughout the performance period, and ends with the formal termination of the contractual relationship. Although the ease or difficulty of contract administration depends in part on the relationships, agreements, and controls created during the formation period, much can be done during the performance phase to ensure project success and reduce commercial risks.

CHAPTER 11

MOBILIZATION AND COMMENCEMENT

Once a construction contract has been awarded, two physical mobilizations begin. The contractor begins to gather the people, material, and equipment it will need to begin work at the site, and the owner should in turn mobilize the personnel, systems, and controls required to manage the contractor's performance. For the contractor, this mobilization requires planning, assignment of personnel, acquisition of hardware such as construction equipment and materials, and the acquisition of funds, bonds, insurance, licenses, and so on. It is an exercise in logistics and, depending on the scope and location of the work, could represent tremendous effort. The owner's mobilization is also a logistical challenge. By the time the first contractor begins site work (ideally before it arrives at the site), the owner should be prepared to perform the twin functions necessary to protect its own contractual interest: technical monitoring and contract administration. This chapter focuses on the initial stages of the latter.

MAJOR ACTIVITIES

Once the contract manager has been selected and given an executed copy of the contract documents, he or she should get acquainted with their contents. Since the job is to ensure contractual compliance of a commercial nature, you need to become intimately familiar with the requirements of both parties, owner as well as contractor. Because it is usually inconvenient to wade through actual contract documents in search of specific requirements—made even more difficult when they are located in several documents or are buried under nonrelated material—most experienced contract administrators prepare contract abstracts. These abstracts reduce the voluminous contract documents to their essential elements—the commercial requirements of both parties. When such compliance aids are used, however, the contract administrator should make absolutely sure they reflect all requirements accurately. In addition, as the contract documents are amended from time to time, such as through change orders, so should be the contract abstracts.

Once the contract manager is familiar with the contract(s) and before the contractor(s) begins substantive work, three initial tasks must be completed:

1. An initial set of contractor submittals must be received, reviewed, and filed.
2. An initial meeting between the contract manager and the contractor's job-site representative is held.
3. The system and facilities for maintaining contract records and conducting contract-related transactions is established.

CONTROLLING CONTRACTOR SUBMITTALS

Most contracts call for the contractor to submit certain items or documents for the owner's review and approval prior to starting certain portions of the work. For some contracts, particularly those involving complex, large scopes of work, these can amount to a mountain of paperwork. To prepare for receipt and review of contractor submittals, the contract manager should review the contract or abstracts and create a list of what is due, and when. Typical submittals required by many contracts include:

1. Evidence of insurance coverage (insurance certificates)
2. Fabrication–installation procedures (rigging methods, weld procedures, concrete placement schedules, soil compaction methods, and so on)
3. Quality assurance or control program documentation
4. Shop drawings
5. Material certificates (such as for stainless steel)
6. Fabrication, shipping, and construction schedules
7. Equipment tagging procedures
8. Equipment operating and maintenance manuals and instructions
9. Signed bond forms
10. Sample products (wall coverings, floor coverings, paint chips, siding samples, and so on)

Although the contract manager should not be qualified or responsible for reviewing the adequacy of all contractor submittals (technical submittals are usually received and controlled by other project personnel), he or she should be responsible for monitoring the status of each and corresponding with the contractor in this regard.

A submittal checklist (Exhibit 9) is helpful for each contract. It should show the contract number, contractor's name, submittals required, date required, and date received and indicate when resubmittal is necessary and resubmittals have been received.

For contracts involving detailed design on the part of the contractor or the contractor's material suppliers, hundreds or even thousands of shop drawings may be required. Quite often, these need weeks or even months to be reviewed, are often rejected and returned to the contractor, and are resubmitted time and time again until finally approved. Without contractual compliance and tracking aids, control of their status becomes impossible. In addition to contract abstracts and log sheets or checklists, use of suspense files (sometimes called "tickler" files) is recommended. These are especially important when controlling the commercial compliance of the owner. Many contractor claims result from the owner allowing submittals to languish unreviewed or undue delay in approving others. The contract manager needs to stay on top of the receipt and acceptance status in order to prevent or defend against these claims—much less to ensure that contractor performance is not delayed through the owner's inability to turn these submittals around within a reasonable time period.

EXHIBIT 9.   Contractor Submittal Checklist

CONTRACTOR SUBMITTAL CHECKLIST

Contract Number _____

Project

Contractor _____

Owner

Submittal	Date Required	Date Received	Resubmittal Required	Resubmittal Received

SPECIAL INSURANCE REQUIREMENTS

Because of the extreme exposure caused by a contractor or subcontractor operating without insurance coverage, or with inadequate coverage, specific cautions regarding verification of this coverage are in order. The contract administrator must make certain that certificates of insurance are sudmitted by each site contractor prior to work and checked against the contractual requirements. It is surprising how many times these certificates are in error, are unsigned, or do not comply with coverages specified in the contract documents. New insurance certificates are often required, and their submittal should be mandatory.

Signatures on insurance certificates are often made by insurance salespersons, agents, or others who may not be able to bind the actual insurance carriers or even to attest to the existence of coverage itself. These signatures should always be verified as to authority and accuracy. Anyone can sign a certificate—the contract manager should ensure that signatures actually rep-

EXHIBIT 10. Insurance Certificate Log

INSURANCE CERTIFICATE LOG

Project

Owner

Contract No.	Contractor	Subcontractors	Date Certificate Received	Expiration Date

resent the coverage stated. To keep up with insurance coverage throughout the project, use an insurance certificate log (Exhibit 10). Both contractors' and their subcontractors' coverage should be monitored, and make sure that coverage does not expire midway through the performance period. Again, suspense files are helpful in this regard.

One final recommendation dealing with insurance. Because the practices of the insurance industry vary frequently (and because agencies and carriers have been known to fail), and policy limits sometimes do not cover increased costs due to price inflation, insurance provisions in standard or reference contract documents should be reviewed periodically. It is not uncommon for them to become quickly out of date or to bear little resemblance to present insurance practices, and for no one to notice—until it is too late.

MEETING THE CONTRACTOR

It is good practice for the contract manager to schedule an initial meeting with the contractor's representative—his or her counterpart—before any significant work is accomplished or any contract administration transactions take place. This establishes the groundwork for good-faith relations throughout the performance period. And potential confusion or delays may be avoided when processing such items as progress payments and change orders if the procedures for handling such are reviewed by both parties before they occur. Additional procedures that the contract manager might consider reviewing are those to be used for backcharges, verification of cost-reimbursable charges, invoice format, notifications, and contract closeout.

Whether the contractor's representative is called the project manager (contractor's), job superintendent, job manager, or any other title, he or she should be the one responsible for commercial transactions that will take place. He should not be merely the contractor's marketing representative or a home office salesperson. He should be on the site, close to the actual construction performance. Just as the contractor should expect not to have to deal with several owner representatives for common transactions, so should the owner have one individual with whom to deal when commercial matters are involved.

Another good idea is for the contract manager to prepare sample forms, or actual ones, for commonly anticipated transactions—price quotations, change order requests, progress estimates, and so on—and to review these with the contractor's representative. Extra copies should be given a future use. This ensures that owner-designed forms are used wherever possible—greatly streamlining future transactions.

If other owner personnel will be working with the contractor, they should also attend this meeting. This may include the construction manager, resident engineer, and construction specialist who will be monitoring the contractor's technical performance. The agenda might include some of the following topics:

1. *Contractual Authority.* Who may speak for the owner and the contractor, by name, for certain commercial and technical issues.
2. *Contractual Communications and Correspondence.* Numbering systems for letters and other documents, addresses, titles, and so on.
3. *Submittals.* The contract manager may wish to give out copies of abstracts or checklists, providing the contractor understands they do not take the place of or modify the original contract requirements.
4. *Changed Work Processes.* Notifications, change-order quotations, change orders, and so on (see Chapter 13).
5. *Progress Payments.* How progress will be measured and paid for. "Master" progress payment forms should be given to the contractor.
6. *Contract Closeout.* And related terms or practices such as retention, credits for owner-furnished items, and release of liens.
7. *Backcharging Procedure.* How notification will be made, contractor and owner obligations for maintaining cost records, and so on.

ESTABLISHING CONTRACT RECORDS

The importance of accurate, easily retrievable contract records cannot be overemphasized. Although contract filing systems may seem a clerical responsibility, they are too important to leave solely to clerical personnel. Even for the simplest of projects and contracts, the amount of contract-related paperwork needed to execute performance and control contractual compliance is sometimes staggering. Without a tightly structured and disciplined approach to these records, contract administration, and indeed project management, is agonizing for everyone.

Figure 15 presents a recommended contract filing system for typical documents, reports, and records used throughout the formation and administration periods. Although the contract manager should concentrate on control of performance-related material once the contractor begins work, he or she should also have information and records concerning preaward events as well. Here are more suggestions on contract record systems;

1. Avoid redundant or duplicate files. Changing or updating is often required, and this can be uncontrollable if changed documents are located in more than one place.
2. Appoint one person responsible for file maintenance and security. The contract administrator is a likely candidate.
3. Control access to and removal of material. Use sign-in/sign-out cards or other methods. Records often disappear.

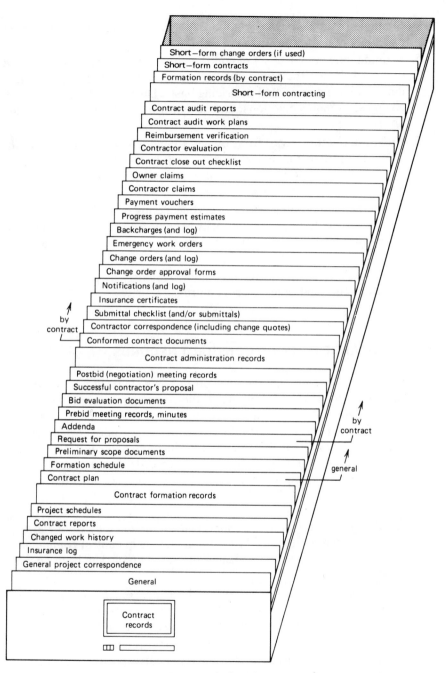

FIGURE 15. Typical contract records.

4. Maintain original contracts in more than one location to guard against loss by fire or other means.

5. Maintain one complete set of contract drawings as they existed at time of award. Resist the temptation to use these for purposes other than determining a contractual or pricing basis. Do not mark them to show changes, as-built conditions, and so on; use other sets for these purposes.

6. Keep contractors away from owner's confidential material. This includes fair price estimates, other bids, and claim-defense documents. In the cramped quarters of many construction offices or trailers, this is difficult to do. Do it anyway.

7. Avoid the temptation to "give everyone a copy of everything." Copying and distribution chores can easily become a nightmare. Periodically review contract-related distribution lists to purge names or positions that have no need to receive copies.

8. Keep track of computer records as meticulously as paper ones. Use security codes and other access controls.

THE IMPORTANCE OF GOOD CONTRACT RECORDS

Contract record systems are often set up in haste, with little planning or regard for their further use. Additionally, the commencement period is usually one in which both parties are in good spirits and an amicable atmosphere prevails. This often lulls owner and contractor alike into neglecting the defensive use of records (or, for that matter, the inoffensive use) for a later, less friendly stage of the relationship. Contract records should be established with a view toward not only expected events but also potential needs, such as:

1. *To Ensure Technical and Commercial Compliance by Both Parties.* Records point out the need for compliance, help control compliance, and demonstrate that compliance, or lack thereof, has been accomplished.

2. *For Contract Auditing Purposes.* The expense, time required, and effectiveness of cost and operational auditing are greatly dependent upon the quality and accessibility of contract records.

3. *For Control of Correspondence.* Contract-related letters, memoranda, telephone conversations, and other correspondence are usually voluminous and may have far-reaching impacts.

4. *Change Control.* To determine the impact of proposed or enacted changes, some means of determining the original and current position must exist.

5. *To Make Progress Payments*. This is particularly true for cost-reimbursable work, where evidence of contract charges are required. It is also important for other pricing methods.

6. *For Quality Control*. Material certificates, erection procedures, test results, and so on, are some of the records needed for this purpose.

7. *To Make Final Payment*. Evidence of completion and acceptability is a prerequisite for contract closeout and final payment. Sometimes it represents events or work that occurred months or even years ago.

8. *To Operate the Completed Facility*. In their zeal to begin and complete the construction process, owners sometimes fail to remember the purpose of the completed facility and neglect operating or maintenance needs. Typical records required include operating and maintenance manuals, equipment wiring diagrams, as-built drawings, warranties, and training material.

9. *As a Claims Defense*. An old maxim holds that he or she who is best documented often prevails in the claims arena.

10. *To Determine What Transpired in the Past*. Because projects can take a long time to construct and many people who are involved at the beginning are long gone by the time it is completed, records represent the collective memory of what transpired. Experience and understandings not captured by records can be lost.

CHAPTER 12

PROGRESS BILLINGS
AND PAYMENTS

Periodic progress payments are a way of life in the construction industry. Unless the duration of a particular contract is shorter than 30 days, intermediate determinations of progress will be necessary. Prompt and accurate contractor invoices and timely payments are critical to the financial status of most contractors. The cost of financing labor, material, and equipment during the construction period is generally too severe to expect a contractor to bear it alone. In addition, payment to its material suppliers and subcontractors is predicated on progress payments to the contractor by the owner.

Contractor concerns in this area are easily seen—they wish payment of the most money in the shortest possible time. There are also significant concerns and risks for owners. Because payment represents the single most powerful compliance lever, owners should ensure that progress payments are determined objectively and are designed to promote timely performance.

Early payment and overpayment for contract work as it proceeds are two risks that should be avoided by the owner. Cash-flow considerations and the high cost of financing dictate that owners should pay only for what has actually been accomplished and not before it has been accomplished. On the other hand, owners who unreasonably delay or withhold earned payments run the risk of contractor dissatisfaction, which can be manifested in reduction of quality, decrease in working forces and equipment, or eventual bankruptcy. All have a negative impact on the project.

ALTERNATIVE BASES FOR PARTIAL PAYMENTS

There are three general bases upon which partial payments may be made: (1) cost, (2) time, and (3) actual performance, or progress. Of the three, payments based on progress are by far the most preferred from the owner's point of view. Examples of each are given in the following paragraphs.

1. *Payments Based on Cost.* The owner contracts for fabrication, delivery, and installation of a turbine generator. The payment method is cost plus fixed fee. Estimated costs and fee total $20 million. Each month the contractor submits invoices representing actual costs for material, labor, equipment, overhead, and so on, incurred during the previous month. After verification for accuracy and allowability, the owner reimburses the contractor for these costs. The fee portion of the contract price is paid in monthly installments or as a lump sum at contract closeout.

2. *Payments Based on Time.* The same turbine generator is bought for a lump-sum price (for furnishing and erection combined) of $20 million. The contractor promises to complete work within 20 months from commencement. A payment schedule is created calling for 20 equal monthly payments of $1 million each.

3. *Payments Based on Progress.* The same turbine generator is bought for a lump-sum price of $20 million, but a progress payment schedule is created with several payment milestones, each weighted in proportion to its respective value. The total value of all intermediate milestones is equal to the lump-sum price. Payment for each milestone is made upon achievement of the actual event in question. Examples of the milestones included are receipt of raw materials, manufacture of turbine rotors, manufacture of turbine stators, fabrication of the generator in the shop, release of the turbine generator for shipment to the construction site, receipt at the site, and intermediate installation milestones, ending with testing and final acceptance.

Of the three methods, partial payments based merely on the passage of time are the least desirable. They offer absolutely no positive incentive for contractor performance. Sometimes time-based payments are proposed where the monthly payment amounts are representative of the "planned" performance during the contract duration; that is, monthly payments are not equal but fluctuate depending upon the amount of work or cost *expected* as the work progresses. Even in these cases, a fixed payment schedule provides no performance incentives. When a payment schedule approximates planned progress, it will probably soon be out of synchronization with actual progress should material shortages, labor disruptions, changes, or other reasons for delay.

Cost-based payments bear a closer relationship to actual progress if one considers that the expenditure of costs is proportional to the amount of work being performed. But even this relationship is indirect and often out of proportion. For large material and equipment items, for example, the cost of the item itself far outweighs the labor cost of having it installed. The incentive is for the contractor to ship the item to the project site quickly, yet little cost incentive remains to have it installed. The incentive is for the contractor to ship the item to the project site quickly, yet little cost incentive remains to have it installed. From the owner's point of view, the majority of progress is achieved upon installation. In addition, contractors interested in increasing their early payments may order or buy as much material and equipment as possible, regardless of when its installation is needed. This loads the site with material long before it is needed, causing storage and protection costs to increase unnecessarily. It is difficult, however, to avoid cost-based progress payments when working under cost-reimbursable contract pricing.

Partial payments are helpful from a control perspective only when they make it beneficial for the contractor to perform segments of the work on time and in a quality fashion, for it should be only then that payment is made for the work in question. One can think of performance-based progress payments as cost incentives to the contractor within the overall framework of risks and incentives determined by the particular contract pricing scheme employed. Even if no absolute incentives are called for in the contract pricing method—as with a firm-fixed-price agreement with no performance, cost, or schedule incentives—*relative* cost incentives may be built into the payment process. The amount of the contractor's payments will not vary (no more or less than the lump sum is paid), but the *timing* of those payments may vary considerably. Because the time value of money is deeply felt by all contractors, timing of progress payments becomes a valuable control tool in the hands of the owner.

Performance-based payment terms are also valuable because they may place significant importance on certain elements of work that, although critical to the owner, have little cost or value in and of themselves. Testing is a good example. Though the contractor's budget for testing may be small, the owner may weight this activity heavily in its progress payment scheme, prompting the contractor to complete testing as soon as possible in order to begin using the finished project or component.

In most cases, the desires of owners and contractors are in exact opposition when weighting the milestones or elements of work. The owner places heavy dollar emphasis on later stages of accomplishment to ensure their completion, whereas the contractor prefers the contract pricing to be slanted toward the front end of the job, increasing its cash flow and reducing financing costs. This desire is the genesis of "front-end loading," of which contractors are often accused when pricing work elements in their bids.

Because they represent the most powerful compliance control, progress-based payments (or progress payments) will be emphasized in this chapter. Since payments under this scheme rely on thorough planning of payment elements and values, the incorporation of specific terms for measurement and payment in the contract documents, and careful administration controls, they should be considered throughout the contract planning and formation periods rather than only during contract administration.

DIFFERENT REASONS FOR MEASURING PROGRESS

There are other reasons, however, for owners to be concerned with contract or project progress. These may include the owner's need to be informed as to the planned completion date, of schedule performance, anticipated project costs, or estimates of the total project upon its completion (forecasts). These are equally important reasons for measuring progress, but in many cases they are based upon criteria different from those embedded in contract payment schemes.

Sophisticated owners of large projects often employ project planning and control methods and systems (often computer based) in order to assess current progress and forecast future progress in terms of time and costs. To the extent that progress measurement for these purposes uses the same milestones, criteria, or weighted values as do contract payment schemes, these two objectives may be met by using the same techniques and data. However, it should be remembered that most owners have two distinct reasons for measuring progress: (1) to determine where the project has been, where it is, and where it is going in terms of cost and schedule; and (2) to facilitate payments to performing contractors in a manner that promotes timely performance of contractual obligations. To force a progress-measurement technique or system designed to accomplish one of these objectives on the needs of the other is not recommended.

THE CHALLENGE OF OBJECTIVITY

One of the most time-consuming, confusing, and often inaccurate ways to measure progress for a construction contract is to make a subjective determination based on the experience of the measurer or the visible achievement at the job site. Many contract documents state that a "mutual determination of progress and commensurate payment will be made at the end of each month," without specifying the criteria for that determination. This usually results in arguments between contractor and owner representatives as to how much has been accomplished and how much it is worth. It is a dangerous practice for both parties—much like gambling—for each could win or lose.

To avoid subjective determinations, objectivity must be planned during the contract formation stage and implemented during the performance period. This requires clear and concise measurement terms in the RFP and contract documents. Contractual scope of work should be broken down into smaller, more discretely priced segments, each readily determined and separately priced. When this "line item" list is inserted into the RFP, contractors can plan their cost performance and bid competitively for each payment item. Once a successful bidder is chosen, the payment items are transferred to the contract documents and form the basis for objective measurement of and payment for progress.

An alternative is to subdivide the work for pricing purposes *after* the contract has been awarded. This calls for the mutual preparation and acceptance of a "schedule of values" that totals the contract price. Although a schedule of values provides for objective progress measurement and payment, these are seldom achieved without the injection of subjectivity by both parties. And because the contract has already been awarded, the contractor is usually in a stronger bargaining position. The benefit of competitively bid payment terms is lost with this approach, and it is not recommended.

Two types of terms regarding progress payments are needed in contract documents. The payment items and their prices should be established; this is generally done in the proposal section of the RFP and subsequently in the agreement section of the contract documents. In addition, specific measurement and payment "terms" are necessary. These describe each payment item and the methods by which it will be measured and paid. These clauses are generally contained in the special conditions, for they vary with each contract according to the scope of work, payment method, and owner control objectives—not to mention the terms that the contractor may insist upon.

Once discrete payment items are identified and weighted as to price and their measurement and payment are defined in the contract documents, objective must be ensured throughout the performance period. This is done through complete understanding of pricing and payment conditions by both parties, especially by those persons who will be requesting progress payments (contractor) and approving or making them (owner). In addition, control and understanding of the impact of contract changes must be accomplished as the work progresses. Specific provisions must be made for pricing and payment of change orders and backcharges.

Of all suggestions made regarding progress payments, none is more significant than this: *Progress measurement and payment terms and practices should reflect the way in which the actual contract work will be performed.* To do otherwise invites difficulty if not disaster throughout the entire payment process. Those responsible for choosing separately priced elements of work, defining milestones for payment, and weighting the value of each should be familiar with the expected method of fabrication, shipment, installation, or erection involved.

Suppose an owner is concerned with timely completion of a three-mile-long buried pipeline. It may decide that pipe should be paid for in sections, say per foot of pipe installed. Payment terms may require acceptance of each foot of pipe before the equivalent installation price is paid. But these two criteria are conflicting, since the contractor cannot gain acceptance of the first foot of installed pipe until the three-mile system is completed and tested for leaks. The pipe will not be installed, connected, backfilled, and tested in one-foot segments! This is an example of progress payment criteria developed in a vacuum, without knowledge of planned construction techniques. It often occurs when personnel responsible for contract formation (the development of payment terms) are not experienced in the other major phase of contract management—administration.

Another example of this problem might involve the installation of concrete in a massive foundation. The contract could call for payment on a "cubic-yard-installed" basis. However, the contractor must expend a great deal of effort and cost before the first cubic yard of concrete is poured. It will have to excavate the entire foundation, construct form-work, and install reinforcing steel and embedments. Each of these steps involves physical progress yet according to the envisioned payment scheme represents no progress. Contractors encountering such preposterous payment schemes should immediately object to their use.

For some contractor activity, particularly that involving services, it is difficult to determine products or to compare output to resources or costs expended. For these particular elements of work, some subjective measurements may have to be made. An example might be guard services provided under contract. There is no way to measure the performance of this service in a quantifiable manner, for the amount of labor expended to guard the facility is not directly proportional to any "products"—so many guards will work for so many hours regardless of the number of thefts or security violations prevented. For these items, cost-based or time-based payments are unavoidable.

SPECIFIC MEASUREMENT TECHNIQUES

Let's examine the most commonly used measurement techniques. Although they have different names, all are similar. They each require the preperformance identification of work steps or products during the formation stage and the assignment of a dollar (or percentage of the total price) value to each. Each of these weighted "accomplishments" should be defined so that field personnel can easily determine whether they have been successfully achieved.

Intermediate Milestones

This is probably the most often used measurement technique, particularly for fixed-price contracts. Each specific portion of the work is broken down into its components so that the completion of a component can be readily identified. Each component ("work item" or "line item") is given a value, and that amount is paid when the item has been accomplished.

Example
A $30,000 contract price for the foundation of a small building may be broken out like this:

Foundation excavation:	$5,000
Concrete footers:	2,000
Base mat:	6,000
South wall:	2,500
East wall:	2,500
North wall:	2,500
West wall:	2,500
Interior supports:	2,000
Ground-level concrete deck:	3,000
Cleaning and dressing and Final acceptance by owner:	2,000
Total:	30,000

Provided that the sequence of construction follows this payment sequence, this may be a proper progress payment approach. Should the sequence of construction call for the simultaneous pouring of all four walls and the base mat, however, another more representative breakdown would give better results. Once the work has started, owner personnel would check the progress at the end of each payment period, determine which of the steps above have been accomplished, and make payment accordingly.

Should any of the preceding steps (or products) be partially complete, the payment terms may not allow partial payment. For example, if at the end of the second month the north wall is 90% complete, no payment of the $2,500 allocated for that wall is made. An alternative approach is to allow partial payments for partially completed milestones. The determination of degree of completion would then revert to a subjective one—for example, is the north wall 90% complete, or is it 80%?

When using the intermediate milestone approach, ensure that the milestones are easily identifiable and that their accomplishment can be achieved in a relatively short period of time. A $50 million contract containing only four intermediate payment milestones (spaced six months apart) would not be practical. Milestones should be selected to allow payment for progress to approximate closely, or follow, the actual accomplishment of progress. Remember that the term *milestone* in this context refers to an identifiable payment item rather than to a significant event in the overall project schedule ("schedule milestone"). They can be the same, but need not be.

Equivalent Units

This technique is also common. It uses performance or product units that are indistinguishable from one another. Each is given the same value, and each should therefore require the same resources for achievement. Examples of equivalent units may be feet of piping installed, cubic yards of earth removed, or tons of structural steel erected. It is easy to see why this technique is tion for unit-price contracts. At the end of each payment period, a determination of the number of units performed is made. The payment price is calculated by multiplying the "unit payment amount" by the quantity of units.

Although lending itself to unit-priced agreements, this technique may also be applied to lump-sum pricing schemes. Lump-sum line items in the contract may be divided into equivalent units, with partial payments of the total lump-sum amount made in proportion to the number of units completed.

Example

A lump-sum contract for $1 million is awarded for construction of an office building. Of this amount, $50,000 is allocated to the foundation excavation, and an estimated 20,000 cubic yards of earth must be removed. Payment is then made on the basis of $2.50 per cubic yard of soil excavated. At the end of the first payment period, measurements show that 8,000 cubic yards have been excavated. Therefore, $20,000 is paid for the partially completed excavation.

Significant problems develop when this technique is applied to units of work that are not exactly equivalent. For example, the equivalently priced unit may be cubic yards of concrete poured. If the scope of work calls for a base mat foundation as well as several elevated concrete decks, the cost and time required to place each cubic yard of concrete in the base mat will be far less than that required to pump each cubic yard of concrete to decking on higher floors. Even though each cubic yard of concrete is indistinguishable from the others—regardless of where it is placed—economy of scale and ease or difficulty of placement will make units installed in one location more costly than those installed in the other. For these situations, separate unit payment amounts should be considered. The same situation arises with piping. It goes without saying that 1-inch-diameter copper tubing should be priced and paid for differently than 10-inch-diameter carbon steel pipe. Again, persons responsible for designing the payment scheme should be knowledgeable in the products, methods to be used, and general construction sequence in order for payment provisions to match actual construction conditions.

A final caution applies to equivalent unit techniques. Should the achievement of the first products (payment units) be accomplished only after a great deal of preparatory effort and expense, a milestone technique should be considered. For concrete placement, a milestone scheme calling for formwork, reinforcement, embedments (anchor bolts, pipe sleeves, plates, and so on), and actual concrete pours may prove more representative of actual work than one calling for payment based only on cubic yards of concrete placed.

The 50–50 Technique

This technique allocates half of the payment amount to commencement of the item in question and the remaining amount to its completion. No intermediate measurements are made; that is, when the item is 90% complete, only the 50% allocated to its commencement is paid. This technique has limited application to major construction efforts. It is used for small items and items for which intermediate progress determinations are difficult or impossible to make. For some payment items, the technique will overestimate actual progress—as soon as the work on the item is started, it is not necessarily half finished. For other items, the technique will underestimate actual progress, as with the case where 90% is finished but only half credit is given. When many small items are measured by this method, chances are that the overestimated ones will balance the underestimated ones to some degree and that a grossly accurate representation of aggregate progress will be made.

When the 50–50 technique is used, care should be taken to ensure that the contractor is not encouraged to start work on all items—in order to capture the initial 50% of their value—with little incentive to complete the first item. In addition, the items in question should be ones that can be accomplished in short periods of time. Otherwise, this technique may give unacceptable results. In general, then, the 50–50 technique is a form of milestone technique, with the start and finish of each item representing milestones of equal weight (50% of the payment amount).

The 0–100 Technique

This is a further adaptation of the milestone technique in which there is only one milestone for each payment item—completion. No payment is made until the item in question has been completed, and full payment is made at that time. Again, as with the 50–50 technique, this method should have limited application and be restricted to minor elements of work that are completed soon after they are begun.

Intermediate or final testing may be examples of the proper use of the 0–100 technique. If the line item in a lump-sum contract for equipment installation allows $500 for testing, this may be paid only upon successful completion of the test involved. It would be difficult to determine any intermediate progress on the test. Installation of minor devices or instruments may also be priced this way. For example, if 100 pressure gauges are to be installed on a complicated piping system, and if the installation of each requires four hours, there is no need to determine partial progress for each gauge. The owner would simply count the number of gauges installed at the end of the payment period and pay for the value assigned to each.

The 0–100 technique alternately may be called the "all or nothing at all" method. Another example of its use is the payment for material or equipment as it arrives at the job site. Of the material cost of each item, 100% is paid as the item is received, and it is meaningless to estimate the partial receipt of a single item—for example, if it is in transit, is it 20% there, 45% there, or 90% there? The owner simply determines if it has arrived or not. If so, the associated price is paid; if not, the material price is not paid.

Subjective Techniques

As mentioned earlier, subjective determinations of accomplishment should be avoided, but sometimes this is not possible. Let's say that the owner has contracted for project scheduling services. This is often called a *level-of-effort contract* because a certain level of effort on the part of the contractor will be expended, it is not generally proportional to the "products" that result, if in fact any identifiable products exist. If the scheduling contract in question is firm fixed price (say, $50,000), partial payments are generally based on elapsed time or cost—in this case, perhaps through prearranged monthly installments, or in some proportion to the number of hours expended during the month for scheduling.

At times, one sees the use of quasi-objective techniques in conjunction with a subjective basis. In the scheduling contract, for example, the owner may ask the scheduling contractor (or consultant) to estimate the percentage of completion of the project scheduling effort. If the contractor replies that it is 40% finished with scheduling duties, a payment of 0.40 of $50,000, or $20,000, may be made. This demonstrates the "mask" of objective measurement applied to the face of subjectivity. For if one investigated the logic used by the contractor to achieve the 40% estimate of completion, it would probably have a cost or time basis. It may have decided that 1,000 hours would be used to perform all scheduling work and that 400 hours have been spent so far—thus the conclusion that 40% of the work has been accomplished. This is a cost-based determination. On the other hand, it may have assumed (or known) that the scheduling effort will take place over 10 months and that its schedulers have been working on the project for 4 months—again, leading to the estimate of 40% completion. This logic is based on elapsed time.

Variations and Combinations

There are thousands of variations and combinations of the five techniques described. For example, milestone techniques may allow the determination of intermediate progress between milestones in a subjective or equivalent-units manner. And there is nothing magic about the 0–100 or 50–50 allocations. These may instead be chosen as 10–90, 25–75, or even 12–23–42–23. When this is done, however, the percentage attributed to each event (or milestone) becomes essentially the same as the weighted value applied to intermediate milestones. As owners and contractors become more familiar with each technique, they soon realize that they are all simply variations on milestone methods. And they are all designed to achieve the same objectives: to identify measurable accomplishments and assign each one a relative payment value, one that's fair, easily measured, and agreed upon in advance.

SELECTING PROGRESS PAYMENT SCHEMES AND
WEIGHTED VALUES

Regardless of who chooses the methodology and values used to make progress payments, the determination should be made as early in the contracting process as possible. The owner should specify the details of progress payments in the RFP and the resulting terms made part of the contract documents. They should certainly be established prior to the first payment period.

Many owners define the methodology to be used and allow contractors to bid the weighted values themselves. Then an examination of the reasonableness of bid amounts is made part of the evaluation process, with any discrepancies isolated for preaward negotiation; again, the owner will typically wish to back-end load, and the contractor to front-end load values. Other owners allow bidders to propose payment schemes and/or relative amounts to each intermediate milestone or unit of measure. Of these two approaches, the first makes more sense. The owner should consciously determine the payment scheme and weighted values; it can do this without knowing bid prices by placing a percentage of the total bid amount in the RFP for each line item.

To determine the payment approach, the owner reviews the scope of work, identifies its control objectives (such as what activity does the owner want to promote), and communicates the details to bidders. Although contractors will object, as they should, to unrealistic methods or values, the owner should take the lead—agreeing to contractor-proposed changes, when they make sense, during the prebid meeting, preaward meeting, or award negotiations.

RECOMMENDED PROCEDURE FOR PROGRESS MEASUREMENT
AND PAYMENT

A typical procedure for periodic (monthly) progress determinations and subsequent payments on a large project is depicted in Figure 16. In addition, the entire payment determination and paperwork flow is shown in Figure 17. Both of these depictions apply to objective measurements of progress, with the contractor making an initial determination subject to prearranged criteria suggested in this chapter.

A basic goal of any payment procedure is that resolutions of progress and commensurate amounts due the contractor should be made prior to receipt of contractor billings, or invoices. Time of invoice receipt should not be the time for both parties to begin deciding how much is earned and owed. A monthly "progress estimating" methodology is recommended, with the results documented on owner-provided progress estimating forms like those shown as Exhibits 11 and 12. These are the tools by which the owner and contractor

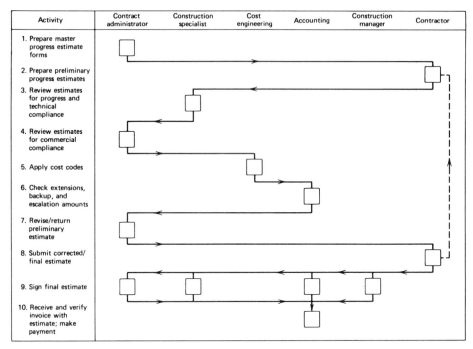

FIGURE 16. Contract progress payment process.

agree and document their agreement. Once an invoice is received and the billing amount agrees with the amount previously determined through the progress estimating exercise, the invoice should be honored.

The format and payment items contained in the estimate forms should be determined at the beginning of the contract administration period and should comply with any contractual payment terms and conditions. The contract administrator furnishes the contractor with "master" payment estimate forms, and the contractor uses these to determine progress each month. It submits completed forms to the owner (contract administrator), and agreement is evidenced by the appropriate signatures. A copy of the signed progress estimate form(s) is sent to the owner's accounts payable group. The contractor then submits a monthly invoice along with the signed estimate form. The accounts payable group receives the invoice, verifies it for accuracy, and compares it with the separately received progress estimate forms (sent by the contract administrator). If no discrepancies are noted, payment is made.

A major premise in this process is that the accounting group is not responsible for verifying performance prior to payment, or for ensuring contractual compliance. This is the duty of the owner's construction staff—contract

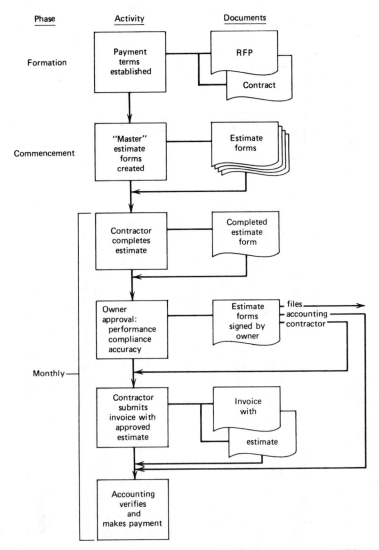

FIGURE 17. Typical paperwork flow: progress payments and billings.

administrator, architect, and so on. If more than one individual "signs off" on progress estimate forms, there is some risk that each one will rely on the other to check certain elements. Each person reviewing estimate forms or invoices should be aware of his or her specific responsibility in the process. As shown in Exhibit 11, each person who signs the forms should be indicating that a certain and specific review has occurred. To avoid disputes, have the contractor sign the form as well.

EXHIBIT 11. Summary of Contract Progress Payment Estimate

SUMMARY OF CONTRACT PROGRESS PAYMENT ESTIMATE

Contract No. _____

Contractor _____

Period Covered:

from _____ to _____

Sheet _____ of _____

(detailed sheets attached)

Owner's Review and Acceptance

1. Compliance with Commercial Terms of Contract

_____
Contract Administrator

2. Compliance with Technical Requirements

_____
Supervising Engineer

3. Costs verified, checked, and balanced; escalation (if any) amounts correct

_____
Construction Accountant

4. Approval to honor invoice if in accordance with this summary and attached detail sheets

_____
Construction Manager

Summary of Progress

Cumulative total as of last period ...... $ _____

Earned this period ...... $ _____

Total earned to date ...... $ _____

Less retention ...... $ _____

Balance ...... $ _____

Less previous payments ...... $ _____

Less Backcharges ...... $ _____

Less other charges ...... $ _____

Amount to be invoiced ...... $ _____

Contractor Attestation

I hereby certify that this Contract Progress Payment Estimate represents the work performed for the period shown above.

_____   _____
contractor's signature   date

EXHIBIT 12. Detailed Contract Progress Payment Estimate

DETAILED CONTRACT PROGRESS PAYMENT ESTIMATE

Contract No. _____
Contractor _____

Period Covered:
from _____ to _____

sheet _____ of _____
Prepared by: _____
Checked by: _____

Payment Item	Description	Unit of Measure	Price	Earned as of Last Period		Earned This Period		Earned to Date	
				Quantity	Amount	Quantity	Amount	Quantity	Amount
PAGE TOTALS:									

FINAL PAYMENTS

The suggested procedure for monthly progress payments should also be followed when the final contract payment is made, with a few additional precautions. More detailed recommendations for final payments are contained in Chapter 17.

RETENTION

A time-honored tradition in the construction industry calls for withholding of earned payment amounts until a later time, when they are released to the contractor. This practice is called *retention*. In general, specified amounts that have been earned through progress on the work are not paid immediately but are held by the owner as an assurance that the remaining work will be performed properly and on time.

Retention is most commonly 10% of amounts due, with release made upon successful completion of the work. It can be determined differently, and it is becoming increasingly popular to establish retention as a certain amount rather than as a percentage (say, $100,000), or to vary the absolute amount or percentage representing retention as the work progresses. To discuss retention more closely, let's examine the reasons for it:

1. *To Motivate the Contractor to Complete the Work.* When used for this reason, retention (or its release) becomes a form of price incentive.
2. *To Cover the Risk of Latent Errors or Omissions.* Should these surface after the contractor has left the site, they may be corrected by using the retained amount.
3. *To Encourage Contractors to Return to the Work After a Planned Demobilization.* Should the value of remaining work constitute a minor price incentive, the contractor may be encouraged to return if completion of that work is a prerequisite to release of a larger retention amount.

The release of retention should vary with the owner's control objectives and the scope of work in question. In addition to the typical release conditions, such as completion of all contract work, retention may be released upon:

1. *Partial Completion of Specified Portions of the Work.* Retention may be released in stages as the work progresses.
2. *Passage of a Specified Period of Time.* The owner may wish to wait 30, 60, or 90 days or more following completion before releasing it to the contractor.

3. *Achievement of Certain Performance.* If operating equipment chinery, or systems are involved, the owner may make certain pe. mance of each a prerequisite to retention release (successful operation, completion of systems testing, after commercial operation, etc.).

Whenever retention is used for any reason, it should be based on actual conditions and perceived risk rather than on tradition or because "we have always retained 10%—doesn't everyone?"

Contractors object to retention in most cases, and rightly so, for it represents their money, the time value of which may be significant. Owners should always realize that it is they who pay this cost—not the contractor. Contractors have to pass this cost along to the owner in the form of higher bid amounts. For large contracts, retention may represent considerable cost to the owner. And owners should recognize the cost of retention and balance it against the value of increased control they may gain through its use.

Unjustified witholding of any monies due the contractor by the owner, including retention, is a poor contracting practice to be avoided. In addition, many jurisdictions place specific restrictions on this practice. When inserting retention terms into contract documents, be sure that they are in accordance with statutory provisions and are enforceable.

ALTERNATIVES TO RETENTION

Consider using one or more of the following if retention is not allowed, does not make sense for the work in question, or is unacceptable to the owner or contractor:

1. *Irrevocable Letters of Credit.* Have a bank or financial institution issue these, to be drawn upon by the owner under certain circumstances. Have the fees and interest charged to the contractor (or owner) only if and when amounts are drawn. This might be thought of as a "retention when needed" approach.
2. *Escrow Accounts.* The owner puts funds in the contractor's escrow account—to draw interest and accrue to the benefit of the contractor—but only released to it upon certain conditions—such as the approval of the owner. These amounts may be the same as they would under retention, or they could be determined according to some other formula.
3. *Back-End Loading of Performance Milestones.* These are just as onerous to the contractor, but easier to keep up with and administer than retention. Also, if retention is not allowed, these serve the same purpose. Some call this "retention in disguise." Remember, the owner pays one way or the other, directly or indirectly, for this practice.

4. *Incremental Withholding With or Without Staggered Release.* You do not have to be bound by the traditional 10% for the entire contract. Consider releasing some of the accrued monies when particular, high-risk work is performed. In other words, try to adjust the withheld amount at all times so that it always reflects the risk associated with remaining work. When retained amounts grow to such an extent that they grossly overcompensate for remaining risk, you are paying for expensive control that you do not need. You are also punishing a contractor who might be in desperate need of the cash—and its subcontractors, who often suffer double retention (like when the owner retains 10% from the contractor, and the contractor in turn retains 10% from its subs).

5. *Other Controls That Achieve the Same Purpose.* These include bonding, insurance, inspection, progress payments, bidder qualification, and the like. Commercial controls can overlap and overkill. Make sure that you've got controls commensurate with risk, and keep in mind that risk varies as the work progresses. Controls cost owners money—be certain that you are getting what you are paying for.

The main message implied by all these suggestions is this: *Be reasonable!* Keep in mind that contractors and their subs live and die by cash flow. Also, check your accounts payable function—make sure that you are not holding invoices for no reason, or "aging" payables to save yourself a few dollars in interest and punishing deserving contractors, vendors, and subs in the bargain. Sometimes the normal delay in owner payments is more onerous than any retention scheme. An old adage here is the "POWER" principle, which goes like this: "*Pay Only When Everything's Right.*" But, to be fair, add this corollary: You only get what you pay for. Owners should remember that they are involved in construction to get a finished facility, one built by competent contractors in business to make a profit. If they have earned it—pay them immediately. You'll reap much greater rewards than if you play games with payments.

REPRESENTATIVE CASES

1. A lump-sum contract totaling $1.5 million for furnishing and installing a concrete chimney and steel liner was awarded to contractor B. Since a lump-sum bid was requested the project purchasing agent saw no need to ask for a price breakdown other than the total lump-sum price submitted by each bidder. At the end of contractor B's first full month of work at the site, a progress payment meeting was held. Contractor B reported $600,000 of work accomplished. The owner's contract manager (the purchasing agent was back at the home office working on the next project) felt that only 10% of the chimney had been installed to date and that contractor B therefore was due

only $150,000. Bitter arguments ensued, with the contractor claiming "front-end" costs of mobilization, material purchases, and so on, that although not incorporated into the chimney were still incurred. The contract manager, seeking a compromise, asked the contractor to prepare a schedule of values for the work and offered that once agreed upon, it would become the objective basis for partial payment. Contractor B prepared and presented the schedule of values. It showed a 90% recoup of the lump-sum price by the third month, even though the contract was to take 10 months to complete. Contractor B refused to change its position and threatened to leave the project if its schedule of values was not accepted. Some years later the contract manager prepared the proposal form for a lump-sum RFP for chimney erection on another project. This is how she structured the bid form:

Item	Material	Labor
A. Excavation for foundation	$ _____	$ _____
B. Concrete foundation, complete	$ _____	$ _____
C. Concrete chimney to elevation 200 ft	$ _____	$ _____
D. Concrete chimney from elev. 200 ft to elev. 400 ft	$ _____	$ _____
E. Concrete chimney from elev. 400 ft to elev. 600 ft	$ _____	$ _____
F. Concrete chimney from elev. 600 ft to elev. 800 ft	$ _____	$ _____
G. Steel liner to elev. 400 ft	$ _____	$ _____
H. Steel liner from elev. 400 ft to elev. 800 ft	$ _____	$ _____
I. Ductwork, complete	$ _____	$ _____
J. Aircraft warning lighting system, complete	$ _____	$ _____
K. Chimney painting, complete	$ _____	$ _____
L. Lightning protection system	$ _____	$ _____
Total material	$ _____	
Total labor		$ _____
Total lump-sum price, including items A–L above	$ _____	

A master progress estimate sheet was prepared containing items A through L, and payment was made as each item was completed. During the course of construction, painting of the chimney was deleted from the scope of work. To make an equitable adjustment in the lump-sum price, the bid (and contract) amount for item K was used to credit the owner for the deletion. This avoided costly and troublesome negotiation over the value of the deleted work.

2. A large pressure vessel for a chemical plant was ordered on a furnish-and-install contract using a lump-sum price structure with provisions for escalation in material cost. The successful bidder quoted a price of $2 million, to be paid in 20 equal installments (it was to take 20 months for fabrication,

shipment, installation, and testing). The owner's contract administrator realized that if this was accepted, the company would be paying on the basis of elapsed calendar time with no regard for the actual progress of the contractor. During the preaward meeting, he proposed a schedule of values that tied the payment of specific amounts to accomplishment of specific progress milestones. The contractor, wanting the work, agreed. Progress payment forms were created as follows:

Delivery of new materials at contractor's fabrication plant	$ _____
Completion of vessel walls	$ _____
Completion of vessel ends (caps)	$ _____
Shipment of vessel and auxiliaries from plant	$ _____
Receipt of vessel and auxiliaries at project site	$ _____
Initial setting of vessel on foundation	$ _____
Installation of auxiliaries	$ _____
Final installation of vessel	$ _____
Completion of hydrostatic test	$ _____
Final acceptance of vessel	$ _____
Total lump-sum price	$ _____

Upon completion of fabrication, the contractor's truck drivers struck for eight months and halted all shipments. No progress payments were made by the owner. Had the contractor's proposed time-oriented payment terms been accepted, a total of eight payments ($800,000) needlessly would have been made, with no progress accomplished during that period.

3. A unit-price contract for furnishing and placing concrete for foundations and elevated decks was awarded contractor C. At the end of the first month, contractor C's records, based on invoices from the concrete batch plant it was using, showed 1,500 cubic yards poured. At a contract price of $40 per cubic yard, it requested payment of $60,000. Suspicious of this tremendous progress, the owner's contract administrator asked the project civil superintendent to calculate the amount of concrete placed by visually inspecting the foundations and reviewing the drawings. From this check the superintendent determined that 900 cubic yards had been placed. Contractor C was told to invoice $36,000 for concrete in place. Furious, the contractor's project manager investigated his operations and found the following:

a. 300 cubic yards had been sold by the batch plant to contractor C but had been used for another nearby project.

b. 200 cubic yards erroneously had been invoiced by the batch plant to contractor C's account but actually had been received by another contractor.

c. 100 cubic yards had been spoiled by contractor C's concrete trucks before ever arriving at the owner's construction site.

4. To ensure that a large cooling tower passed final commercial testing, the design engineer inserted a clause into the specifications calling for a 20%

retention until completion of the testing. As the contract sum amounted to $35 million, and preoperational testing was not to commence for the projec_ until two years after completion of the tower, all seven bidders bitter_ protested this burden at the prebid meeting. They claimed, rightfully, that _ final retention costs would create interest expense for the contractor _ would be passed on to the owner) of 20% of 35,000,000 × 14% int_ compounded annually for two years, or $2,097,200. Cooler heads prev_ and a graduated retention plan was substituted. The new plan called f_ retention during the first year, reduced to 5% during the remainde_ work, and never to exceed $400,000. At the completion of work all p_ due the contractor would be made, with the exception of $200,0_ would be retained until preoperational testing was complete and_ accepted for use.

5. At times, retention is used by the owner as a lever not t_ work is performed in accordance with the contract but to _ contractor who has a scheduled break in performance will act the project site and complete the remainder of the work. In _ schedule of values for a structural steel erection contract all_ erection of steel to be left out in a wall for a turbine buildin_ was to demobilize for six months and return to compl_ penetration after the turbine generator was installed (the_ allow the large equipment to be moved into the building)_ contract was for approximately $4 million, the projec_ $12,000 task might not be sufficient to ensure prompt re_ in time to close the turbine building before the onset of_ $350,000 was withheld after all work except the _ completed. Obviously, the retention far exceeded t_ work, but it did cause the contractor to return and _ as requested. This approach cannot be impleme_ desires; it must be conceived prior to bidding an_ contract documents. To impose retention wit_ bad business practice and could be contrary t_

CHAPTER 13

CHANGE ORDERS

Nothing is more constant than change during the course of a construction project. Despite the best efforts of all concerned during the planning, formation, and administration phases of contracting, changes will almost certainly occur. This chapter describes the typical causes of contractual changes and presents a process for controlling and accommodating changes within the framework of owner–contractor agreements.

TYPES OF CONTRACTUAL CHANGES

Thousands of changes take place during the design and construction of a new facility. Among these are design changes, schedule changes, price and cost changes, resequencing of design and construction activities, material substitutions, and modifications to construction methods, to name only a few. Many of these changes may be accommodated with little problem. In fact, many represent improvements to original plans or refinements made to improve performance when unforeseen events occur or when more information is available concerning anticipated events. In this chapter we will restrict our discussion to those changes that affect contractual agreements: contractual changes. These represent the most damaging of all.

Most construction practitioners categorize contractual changes into two groups: *informal* and *formal* changes. Formal changes are made by the owner and result in written directives to the contractor to change the scope of work, time of performance, prices or just about anything else once the contract is awarded. Formal changes are conscious decisions on the part of the owner to modify the contractor's work.

The most common formal changes concern alternatives to the design of the are manifested in revisions to the construction drawings or speci- ̣n addition, owners often change their schedule needs or rese- ə planned work to accommodate changes in the owners' needs or the perıu̇rmance of other contractors or suppliers. Formal changes have their authority in the changes clause of most contracts. This clause usually gives the owner the unilateral right to change the work and to require the contractor to comply with such changes. The additional compensation, in terms of time, money, or both, necessitated by the change, is determined in a structured, prearranged manner according to the terms of the original agreement (see discussion of pricing change orders in Chapter 3). The written directive ordering the contractor to make the change and specifying additional compensation (if any) is called a *change order*.

Because formal changes are generally identified before they are incorporated, based on a planned and deliberate choice by the owner, and documented by a formal modification instrument—change order—they are usually easy to handle and accommodate. Strict controls are required to assure the owner that these types of changes are identified as early as possible, thoroughly evaluated in terms of cost, schedule, and technical impact on the work, priced economically, and enacted in accordance with the contract provisions. Although the number and frequency of formal change orders is sometimes staggering, most contract administration programs are able to accommodate them with a minimum level of difficulty. The same cannot be said for the other category of change: *informal changes*.

Informal changes are sometimes called "constructive changes," in a legal sense. They represent modifications to the contractor's scope of work or method of performance that result from (1) acts or omissions of the owner; (2) acts or omissions of third parties, such as other contractors, material suppliers, or God; or (3) other forces beyond the contractor's control. They are the most difficult to identify and control. They cause the contractor to perform work differently or to perform different work from that which is required by the contract or which could be normally anticipated. The resulting costs and disruption are the same as if a formal change order had been issued. Once informal changes are identified and resolution between owner and contractor is achieved, they should be transformed into formal change orders.

Informal changes are so troublesome because they originate everywhere, are often identified after the fact, and their cost and schedule impacts are difficult to quantify. Informal changes can also detract from the performance of *unchanged* work. Informal changes are difficult to resolve and easily fester into disputes and claims.

TYPICAL CAUSES OF CHANGE

There are countless causes of change that may impact a contractor's work on such a complex, highly interdependent, and time-sensitive effort as a construction project. Students of these phenomena, however, suggest that a large percentage of changes is caused by one or more of these reasons:

1. *Defective or Incomplete Design Information.* Owners dissatisfied with the products of their engineering departments or the performance of their design consultants would probably vote for this as a major cause. Symptoms of this problem are numerous or massive revisions to design drawings and specifications. Revisions to design—made when the contractor has already commenced work—are not bid competitively. And they often create a vicious cycle: revisions to one system, structure, and so on, causing revisions to interrelated systems, structures, and so on. It is a common phenomenon on fast-tracked projects, where design is often only one step ahead (or behind) construction. It is also a symptom of owners or their designers who do not know what they want or how seriously their indecisiveness affects construction.

2. *Late or Defective Owner-Furnished Material and Equipment.* Many owners purchase equipment or material and furnish it to the construction contractor(s) for erection or installation. Owners who decide to save money or exercise direct control by buying these items themselves must consider the additional responsibility and liability they assume by doing so. The decision to buy directly should be based on owner and contractor purchasing strengths and controls, as well as the economies involved. The question of procurement responsibility should hinge not only upon anticipated price but also on a determination of which party is best equipped to handle the additional responsibility—who can ensure that the materials or equipment will arrive at the site on time and in the proper condition.

3. *Changes in Requirements.* Changes in operating, safety, environmental, market, feasibility, funding, or regulatory requirements, particularly for projects that span several years or concern highly complex or sensitive efforts represent major causes of change, both formal and informal. There is little an owner or contractor can do about these types of changes. At best, every effort should be made to anticipate changes of this nature and provide for their impacts in project budgets and contractual terms and conditions.

4. *Changed or Unknown Site Conditions.* Subsurface conditions represent a classic example of site conditions that may not be known or may change during the course of construction. The occurrence of subsurface water, rock, or other material often directly affects a contractor's performance. Other site changes, though not as common, also occur. These include weather, access, congestion in the workplace, and operating space, such as material laydown areas, warehouse storage, and security restrictions.

5. *Impact of Collateral Work by Others.* When actions of a contractor interfere with, delay, or cause hardship on other contractors, informal changes may result. This risk is intensified in direct proportion to the number, proximity, and interdependency of contractors (and subs) at the site.

6. *Ambiguous Contract Language and Contract Interpretations.* Extensive contract formation controls, consistent and clear documents, and contract administration training and procedures help prevent changes due to this cause.

7. *Restrictions in Work Method.* There are countless examples of owners restricting or changing contractor work methods. Unless it is specified otherwise in the contract documents, contractors are generally free to construct in the manner they deem suitable. Owners who impose restrictions on acceptable or traditional methods, absent any specific restrictions in the contract, interfere with the contractor's performance. As owners become increasingly involved with the daily operations of contractors—crossing the fine line between performance monitoring and supervision—interference becomes possible. As independent contractors responsible for the means and methods employed to satisfy the contract requirements, contractors will claim informal changes if this occurs. And those claims are often successful.

8. *Late or Inadequate Contractual Compliance on the Part of the Owner.* Construction contracts are a one-way street, requiring compliance by both parties. Although much has been said about contractor compliance, owners need to fulfill their contractual obligations as well. Many owners feel that their obligations are restricted to payment of contractor charges. This is absolutely not true! Virtually every contract requires (expressly or implicitly) actions on the part of the owner in addition to payment of the bill. Most of these deal with prompt and timely owner inspection and acceptance of the work product as it progresses. Owners are often required to review design or shop drawings produced by the contractor in a timely manner. This is usually a duty that must be performed within so many days of receipt of these and other contractor submittals. It is not uncommon for owners, in their zeal to manage or monitor contractor performance, to overlook commensurate performance required on their part. The capability of the owner to meet these contractual commitments should be analyzed before they are inserted into contract documents. If you cannot meet these obligations, do not take them on.

9. *Delay or Acceleration.* When a contractor's work is delayed, by the owner, other contractors, or third parties, it incurs additional cost and it might be unable to meet contractual time commitments. The same is true when work is accelerated, or compressed. Typical costs for delays include standby and remobilization expenses, loss of efficiency, and changes in the weather or other conditions. Acceleration costs often involve the increased charges incurred for shift time and overtime, additional work forces and equipment, and loss of efficiency. Both acceleration and delay situations illustrate that a combination of causes and effects, which in turn become causes for further effects, is often found among constructive or informal changes. Separating each, and quantifying their respective impacts in terms of time and money, is very difficult. This is one reason why timely identification and resolution of changed conditions is critical. The alternative is often

litigation or prolonged negotiation. Specific discussion of delay and acceleration claims is contained in Chapter 15.

THE CHANGED WORK PROCESS

The truest test of contract management is the ability to respond to and accommodate changes. Whether they are formal or informal changes, contract administration controls and procedures allow the owner to perform each of the steps described in the following paragraphs in the changed work process.

1. *Identification.* It is surprising how many changes, particularly informal ones, go unrecognized until it is too late to control their impact. Changes should be identified as such as soon as possible in order to allow a thorough evaluation of their impact and a conscious, well-informed decision as to whether and how they will be enacted. Planned or potential changes, when initiated by the owner, should be transmitted to the contractor immediately. This should be done through use of a form letter similar to the notification instrument shown as Exhibit 13. As the term implies, *notifications* are merely instruments by which the owner notifies a contractor of a planned or potential change. They also serve as a request for the contractor to determine the impacts on time and schedule, if any, that the change would cause and to quote on both to the owner by a specified time. Notifications (sometimes called "bulletins") differ from change orders in that they do not necessarily instruct the contractor to perform the changed work. They merely request the contractor's reaction to proposed changes. When designed and used properly, they should:

Inform the contractor of all details of the proposed change. If revised drawings or specifications are involved, these should be listed and attached.

Indicate whether the information constitutes a change, or is merely additional or promised information that does not affect change or does not deviate from the original agreement.

Request a contractor quotation, in a specified pricing manner (lump sum, unit prices, and so on) by a specified date.

Contain a clear statement as to whether the work should commence immediately or the contractor should ignore the change until a change order is issued.

Be signed and dated by the owner. Numbering notifications is also suggested for ease of reference. There can easily be hundreds of these.

Require the contractor to acknowledge receipt of the notification by returning a signed copy thereof. This indicates not that the contractor agrees

EXHIBIT 13. Notification

NOTIFICATION

Date: _____

Notification

Number: _____

To: _____   Contract No. _____

_____   Contract _____

_____

The enclosed ☐ information ☐ document(s) ☐ drawing(s)
☐ are proposed ☐ have been approved
for use on the above contract work.

No.	Description	Identification	Remarks

(attach continuation sheets if necessary)

Please review the enclosed and proceed in accordance with the instructions indicated below.

INSTRUCTIONS

☐ No change required in contract, proceed with affected work.
☐ For information only.
☐ Change in contract work, submit proposal, including all impacts on cost and time of performance, as requested by _____.
(date)

    ☐ lump sum. Do not proceed with work.
    ☐ contract unit prices. Submit estimated quantities and proceed with work.
    ☐ unit prices not included in contract. Submit prices and estimated quantities. Do not proceed with work.
    ☐ cost plus in accordance with contract provisions. Submit estimated manhours and material costs. Do not proceed with work.
    ☐ other: _____

Issued by: _____ date _____
Receipt acknowledged by: _____ date _____
Return one copy of this form, with receipt acknowledged by signature of contractor, to: _____ by _____.
(date)

_____ continuation sheets are attached.

to perform the affected work but merely that the contractor is aware that the change has been proposed.

Be sent to *all* contractors who may conceivably be affected, even indirectly, by the change.

The notification process fits situations where changes are proposed by the owner, so it is well suited for most formal changes. However, many changes are proposed or identified by the contractor. This is often the case with the other category of changes—informal or constructive changes. When this occurs, contractors should notify the owner, identify the change and describe the actual or anticipated impact on cost, schedule, or technical performance. Contractors generally do this by a change request or a claim.

2. *Evaluation.* Once a change has been identified, by either the owner or contractor, the owner must decide whether to adopt the change. If the change is discovered after the fact, its impact must be estimated or determined. The evaluation of proposed or enacted changes involves a review of the contractor's quote (or claim) for accuracy, compliance with any contractual provisions, and reasonableness. Just as fair price estimates help owners evaluate original contract bids, they help evaluate change quotations. Consider using them for this purpose. You might use a form similar to the "Request for a Fair Price Estimate," Exhibit 14. Make sure that the required pricing method is the same as the one used by the contractor for ease of comparison.

3. *Approval.* Often owners go back to contractors requesting additional information or rejecting quotations and asking for a revised quote. Sometimes negotiation is required. When both parties have agreed to the need and amount of additional compensation, a change order is issued. The documentation and approval of the change identification and evaluation process is done through a form similar to the change order approval form shown as Exhibit 15. This form documents the evaluation and justifies the incorporation of the change into the contractual agreement.

4. *Incorporation.* Once a change order is approved by the owner, it is issued to the contractor to modify the terms of the original agreement. It may incorporate the changes of one claim, notification, or contractor change request or, as is often the case, consolidate the collective results of several. Change orders should be numbered and dated and refer to the changes clause of the contract documents. They should thoroughly describe the change and the revision to contract price or time of performance. Change orders should also be signed by the contractor, indicating its acceptance of the changed work as well as the change in compensation. An example change order form is shown as Exhibit 16.

5. *Payment.* Payment for change order work should follow the same procedure as described for normal contract progress payments. The exception is that changed work should be identified and listed separately on progress estimates and invoices. It is strongly recommended that original and changed work not be commingled during this or any other step of the changed work process. Keep track of each separately.

EXHIBIT 14.   Request for Fair Price Estimate

REQUEST FOR FAIR PRICE ESTIMATE

TO: _____   Contract No. _____
FROM: _____   Contract: _____

Please prepare a FAIR PRICE estimate based on ☐ the attached, or
                                                ☐ Notification No. _____

and return by _____.
                    (due date)

Required Pricing Method:

    ☐ Lump Sum
    ☐ Unit Prices (with extensions)
    ☐ Cost Plus (with estimated cost)
    ☐ Contract Unit Prices ☐ Apply, ☐ Do Not Apply
    ☐ Contract Cost Plus Markups ☐ Apply, ☐ Do Not Apply

Additional Instructions: _____
_____
_____
_____

Attachments: _____
                _____
                _____
                _____

                     _____
                       (contract administrator)

These steps are shown in Figure 18. This figure assumes an owner–construction management organization of some size, shows typical participants responsible for the process, and the general flow of paperwork involved. Because control and documentation of changes represent critical challenges during the performance period, and many owner programs are weak in this area, carefully designed documents and records like these are particularly important.

CHANGED WORK DOCUMENTS

Generic examples of documents to control and record the changed work process are included as exhibits in this chapter and shown in relation to one

---

EXHIBIT 15. Change Order Approval Form

CHANGE ORDER APPROVAL FORM

Number _____

	Source of Change	Date

Contract No. _____
Contract: _____

- ☐ Notification No. _____ _____
- ☐ Contractor Claim _____
- ☐ Contractor Request _____
- ☐ Other _____ _____

Pricing Method to be Used: _____

Estimated (or Lump-Sum) Price: _____

Cost Evaluation

Item No.	Description	Pricing Method	Quantity	Fair Price Estimate	Contractor Quote

Schedule or other impacts: _____

_____

Prepared by: _____ _____
(contract administrator) (date)

Reviewed by: _____ _____
(construction specialist) (date)

☐ Approved ☐ Rejected by: _____ _____
(construction manager) (date)

If approved, list Change Order number(s) to be issued: _____

_____

_____

Additional Disposition Instructions: _____

_____

_____

_____

EXHIBIT 16.   Change Order

CHANGE ORDER

Contract No. _____   Change Order No. _____

Contract _____   Project _____

To: _____
_____
_____

In accordance with Paragraph _____ of the General Conditions of the above named contract, the following change is made and incorporated into said contract:

_____
_____
_____
_____

The Contract Price is changed as follows: _____
_____

The Contract Performance Dates and/or Durations are changed as follows: _
_____

All other terms and conditions of subject contract remain in full force and effect.

Prepared by: _____   _____
    (contract administrator)   (date)

Approved by: _____   _____
    (construction manager)   (date)

Accepted by: _____   _____
    (contractor)   (date)

Please sign and return one copy to: _____

FOR OWNER USE ONLY

Original Contract Price: _____
Current Contract Price: _____
Notification No.   _____
Change Order Approval
Form No.   _____

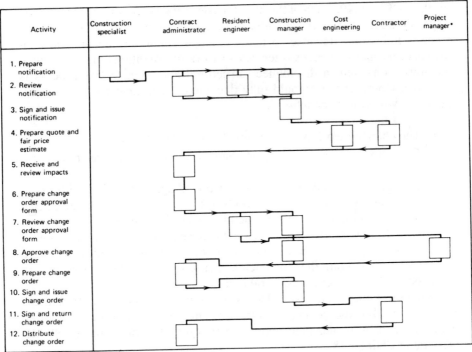

FIGURE 18. Change order process for a large project organization.

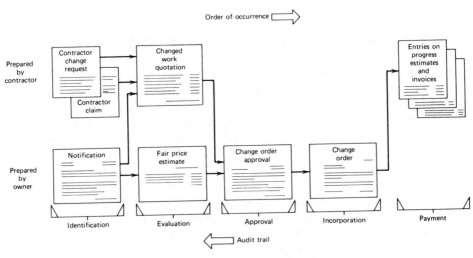

FIGURE 19. Changed work documents.

another in Figure 19. Each of the five steps described earlier is represented by a specific document or documents, and their recommended order of occurrence is indicated. Because so many changes commonly occur, and considering that many changes often revise work or conditions that have been previously changed, a disciplined, structured approach to this process is essential. Contract files should fully document the changed work process. This has two primary benefits:

1. The changed work process is much more manageable.
2. Operational or cost auditing of changed work is more straightforward.

The changed work process described in this chapter, together with documents depicted in Figure 19, meet both of these objectives. The audit trail provided by a complete ''set'' of documents used during each step is clear and easy to follow. Notice that each document refers on its face to the number or other identity of associated documents that precede or follow. Without these ties, or without the documents themselves, control and review of changed work becomes an unbearable task.

Not only should each changed work incidence by carefully controlled and documented, but the *cumulative effect* of changes should also be reported and analyzed by owner management. This type of effort is more fully described in Chapter 18.

Other compliance controls ensure that changes are being handled properly, such as the use of a notification log to record and track the status of notifications. An example is shown as Exhibit 17. A change order log (Exhibit 18) and a changed work history form (Exhibit 19) also help. They are extremely useful as checks against change orders identified on progress estimates and/or contractor invoices.

EXCEPTIONS TO THE RECOMMENDED CHANGED WORK PROCESS

Occasionally (or all too frequently), you will not have time for a formal process such as described in this chapter. Emergencies or other unusual conditions demand more expedient means, and often these involve changed work. When this occurs, contractors are usually instructed (sometimes verbally) to proceed with the changed or affected work immediately, with resolution of its costs at a later time.

Those responsible for managing construction efforts on the part of the owner should have the authority to accommodate such emergencies or unforeseen conditions as rapidly as possible. By the same token, however, this exception should not become the rule—emergency authorizations should not be used excessively or to implement changed work that could be processed under more controlled methods.

Simply because standard change-control processes cannot be imple-

EXHIBIT 17. Notification Log

NOTIFICATION LOG

Page ___ ___

Project: _____
Owner: _____

Notifica-tion Number	Contract Number	Contractor	Date Issued	Date Receipt Acknowl-edged	Remarks

mented for emergency work does not mean that all control should be abandoned. When this work takes place, it is usually performed in a cost-reimbursable manner. This is understandable, as the scope and cost of such work is usually impossible to quantify before proceeding. However, the same types of controls recommended for cost-reimbursable contracting should be implemented. Some owners issue "emergency work orders," or other change-oriented instruments to cover this type of occurrence. Time sheets, material receipts, and other evidences of cost should also be collected on a

EXHIBIT 18. Change Order Log

CHANGE ORDER LOG

Page _____

Project: _____   Contract No. _____
Owner: _____   Contractor: _____

Change Order Number	Descrip-tion	Notifi-cation Number	Pricing Method	Estimated or Lump-Sum Price	Date	
					Issued	Copy Re-ceived

EXHIBIT 19. Changed Work History

CHANGED WORK HISTORY

CONTRACT NO. _____   CONTRACTOR _____

Notification No.	Description	Issued Date		Ac-know.	Proposal/Estimate Submitted					Proposal/Estimate Resubmitted					Resolved					Final Cost	Comments
		To Contr.	To Est.	Recv'd	Contractor		Estimating		Contrac-tor's Quote	Contractor		Estimating		Re-jected	Change Order					
					Sched.	Act.	Sched.	Act.		Sched.	Act.	Sched.	Act.		No.	Issued	Accept.			

daily basis. When the work is completed, or proceeds to the point where either the emergency is past or a reasonable estimate of the cost may be established, issue a standard change order to cover it so that *all* changed work is eventually documented and controlled in essentially the same manner. Because changed work represents one of the primary areas of commercial risk, we will return to it in other chapters.

REPRESENTATIVE CASES

1. A unit-price contract for furnishing and installing structural and miscellaneous steel contained an extensive price list for virtually every category of item that could be needed for a coal-fired power plant project. Of the several hundred priced items, one was for furnishing and installing "metal building frames and louvers" contemplated for the turbine building air exhaust system at a price of $40 per pound of steel. After 18 months of work, the design was changed to require the addition of eight "exhaust vents with adjustable vanes." A change order was issued upon a lump-sum quotation from the contractor of $88,000. While reviewing the change order files, an independent auditor later pointed out that the additional "exhaust vents with adjustable vanes" were identical to the "metal building frames and louvers" for which unit prices existed in the contract. Since each vent weighed approximately 200 pounds, the use of the existing unit prices would have cost the owner 8 × 200 × $40, or $64,000, rather than the $88,000 authorized by the change order.

2. In order to maintain budget control, the project manager adopted a single limit of $25,000 for individual change orders issued in the field by the on-site construction manager. A copy of each field-issued change order was reviewed by the home office to ensure that this policy was followed. A complete review of all change orders took place after the project was completed and all contracts closed out. At that time, it was discovered that the site manager never issued individual change orders exceeding $25,000 for the same work. For example, the rerouting of buried pipeline called for $72,000 in extra work. The construction manager divided the work into three separate change orders, each issued several days apart and interspersed with other nonrelated change orders. The change orders in question were:

a. C.O. 109, reroute buried piping, $24,500, issued July 19

b. C.O. 113, reroute buried piping, $24,000, issued July 23

c. C.O. 121, reroute buried piping, $23,500, issued July 26

The construction manager had in effect issued a $72,000 change order in three installments all within one week—contrary to the spirit, if not letter, of the project manager's policy.

3. A lump-sum contract for structural steel erection contained unit prices for potential additions or deletions to the scope of work. For a particular size of wide flanged steel members (W12 × 115) the add price was $1.15 per

pound, and the delete price was $0.85 per pound. Before the contractor ever began work on building 2, a flood of drawing revisions was issued. Among many other changes, the drawings alternately deleted and added steel members of the size W12 × 15 as follows:

Revision Batch 1: delete 985 pounds W12 × 15 (March 15)
Revision Batch 3: add 1,785 pounds W12 × 15 (April 30)
Revision Batch 4: delete 640 pounds W12 × 15 (May 28)
Revision Batch 7: add 1,800 pounds W12 × 15 (July 14)

A change order was issued following each notification to adjust the lump-sum price according to the additions or deletions in question. The cumulative effect of the four change orders in question was $5,504:

Delete 985 pounds (0.85) = $ 837.25
Add 1,785 pounds (1.15) =   2,052.75
Delete 640 pounds (0.85) =   544.00
Add 1,800 pounds (1.15) =   2,070.00
    Total $5,504.00

In actuality, these were merely "paper" changes, as no work had begun on the affected areas. The contractor was overcompensated because the *net* effect of the additions and deletions (be it a reduction or increase in quantity) is the only basis for an adjustment in contract price. In this case, the net effect was:

(985)
1,785
(640)
1,800

1,960 pounds net addition

And the owner should have issued one change order covering this net addition in the amount of 1,960 × 1.15 = $2,254, rather than the $5,504 authorized by the four separate change orders.

4. The owner of a petroleum tank farm some 2,000 miles from its home office, where design and project management took place, implemented a policy calling for the use of two distinct change order types: (1) home office change orders issued for changes due to revisions to home office–issued drawings and specifications, and (2) field office change orders issued at the construction site for site-related changes. To ensure that change orders were not issued by both sources to cover the same activity, procedure called for duplicate copies of each field change order to be sent to the contracts department at the home office. The procedure served its purpose well, until the field contract administrator fell behind in copying and mailing chores and became lax in describing the cause and nature of the work on the field change-order forms. As a result, the home office was not adequately informed of the field

orders and consequently issued change orders of its own, resulting in duplicate or overlapping payments to the contractor that went undiscovered.

One of these instances involved relocation of underground utilities. The home office realized from existing drawings of an adjacent plant that gas, water, and electrical lines would have to be relocated to make room for a new tank foundation. The quoted price and change order included appropriate compensation. By the time the home office change order had been issued, however, the contractor had begun the relocation work and had requested and received a field change order for the relocation from the contract manager, who assumed the underground utilities were a site-generated occurrence. His description of the work stated simply "extra labor and material for tank foundation installation." Not realizing this actually meant utility relocation, the duplication went unnoticed in the home office when they finally received a copy of the field change order months later. The contractor, having been paid twice for this work, was long gone.

5. To expedite relocation of the main access road to a pulp and paper mill under construction (floods had destroyed the existing bridge approaching the site), the owner's contract manager issued a cost-plus change order to the general contractor based on a hasty estimate of $50,000. A copy of the order was sent to the home office accounting section, and the next invoice from the contractor included the change order number, date, and $50,000 as the amount. It was paid and forgotten. As it turned out, the actual cost to the contractor amounted to $26,500, and with a contractual markup on changed work of 20%, the change order should have been rescinded by one based on actual costs as soon as they could be ascertained (in this case, $31,800). Had a cost-plus work order been issued for an estimated amount of $50,000 and the accounting section instructed not to honor invoices based on any other instrument than change orders, this overpayment of $18,200 would have been avoided.

CHAPTER 14

BACKCHARGES

Sometimes it is necessary for the owner to perform certain work operations, or have them performed by others, when a contractor or supplier is unable or unwilling to do the work it is supposed to under the contract. When these situations occur, the cost of performing the work should be charged to the offending contractor or supplier in the form of a backcharge.

COST COLLECTION

Other than administrative time and expense, the net cost effect on the owner resulting from a backcharge incident should be zero. That is, the additional cost to the owner for having the work performed by another contractor (or doing it itself) should equal the deduction from payments due the offending contractor. Some owners pay for the additional work and later invoice the offending contractor for reimbursement. This is not recommended, as it places the owner in a defensive position—the burden (and risk) of collection is unnecessarily placed on the owner. To avoid this, the owner should deduct the backcharge amount from payments due the contractor by either a negative entry on monthly progress estimates or a deduction (credit) to the contractor's invoice. In cases where the backcharge situation is discovered only after the offending contractor has left the site and been paid in full, the owner has little choice but to invoice it and pursue collection. Whether to add a markup to the backcharge amount to cover the owner's "handling fee' is another issue. Usually, however, the nuisance value and ill feelings this creates far outweigh the minor amount to be charged.

BACKCHARGE SITUATIONS

To understand the role of backcharging in a construction environment, let's look at some typical situations where a backcharge is in order.

Example 1

Contractor A has been awarded a contract for excavation and pouring concrete building foundations. Included in the scope of work are placing and installing steel anchor bolts in the base slab of a large structural steel building. When this slab was poured, the contractor neglected to include certain bolts that were added by a recent drawing revision and are required to secure several major structural steel column base plates. The structural steel erection contractor (contractor B) arrives at the site and discovers the omission immediately. Contractor A has demobilized and left the site and will not return for several months to work on other foundations. Contractor B, therefore, is issued a change order to correct the deficiency. Contractor A should be backcharged the amount of that change order.

Example 2

Contractor C, a piping installation contractor, is receiving and unloading a shipment of owner-furnished prefabricated pipe when it realizes that extensive weld preparation of the ends of each spool piece has not been performed. The weld preparation is specified by the purchase order for pipe from vendor D. Vendor C is notified, admits the error, and suggests that the pipe be shipped back to its plant and corrected. Vendor D expects to have the pipe returned to the site in six months. Rather than delay contractor C for this period, vendor D agrees to pay for the rework if done in the field by contractor C. Vendor D signs a backcharge form, a change order is issued to contractor C to perform the weld preparation, and the amount is deducted from vendor D's billing.

Example 3

An employee of contractor E, suffering from an extreme hangover, mistakenly drives a bulldozer over a carton of sensitive (and fragile) electrical instruments waiting to be installed by an electrical contractor, F. Contractor F asserts that it has no contract or legal relationship with contractor E and demands a change order from the owner covering the replacement cost of the damaged items. Contractor E agrees to be backcharged for the amount in question.

Example 4

While starting up a large pump to test a completed fuel oil system, the owner's engineer hears a piercing sound and immediately shuts the motor off. On inspecting the motor, he finds the piping contractor (contractor C) carelessly left a bundle of welding rods inside the pump's intake piping. The pump is destroyed and a replacement ordered. Contractor C has long ago left the site and received final payment for its contract. A mechanical subcontractor on site agrees to install the replacement pump, and a backcharge to contractor C for the cost of the pump as well as the installation work is issued. An invoice for the total cost is sent to contractor C.

SUGGESTED BACKCHARGE PROCEDURE

The contract manager should be notified of any backcharge work before it starts. The manager determines the scope of the work to be performed based on discussions with project personnel, correspondence, photographs, and so on, and requests an estimate from the party who will be performing the backcharge work. The cost estimate may be submitted on the basis of a lump-sum price, unit prices, including estimated quantities, or cost plus with an estimated total cost.

A backcharge form (Exhibit 20) is completed with the scope of work and total estimated (or actual) cost. The backcharge form, including any applicable backup information such as details of the cost estimate, correspondence, and telephone or meeting memoranda, is forwarded to the offending contractor or supplier (for whom the work is to be performed) for signing. This signature should be obtained prior to commencement of the backcharge work if possible. If the offending contractor is a material or equipment vendor, its representative is immediately called to the site to inspect the product and sign the backcharge form. Copies of the backcharge form are distributed to appropriate project personnel, including accounting, who are responsible for invoicing the contractor or supplier upon completion of the backcharge work (or periodically, depending on the duration) or crediting contractor invoices when possible. A recommended process for backcharging is shown by Figure 20.

REPRESENTATIVE CASES

1. A vendor of heating, ventilating, and air conditioning equipment was to ship several large fans with associated motors to a heavy industrial construction project. When the shipment arrived, the installation contractor noticed that 32 motors accompanying the fans were the wrong size and model; they were incompatible with the fans and not in conformance with the purchase order specifications. The construction manager ordered the installation contractor to buy the proper motors from its own sources and begin installation. A cost-plus change order was issued the contractor for its invoiced amount of $16,000 plus $1,600 for profit and overhead. When a backcharge of $17,600 was sent to the fan vendor, it refused acceptance of the charges, claiming that the mistake was not brought to its attention when discovered. After checking,

EXHIBIT 20.   Backcharge

BACKCHARGE

Contract/Purchase Order Number _____   Backcharge No. _____

Contractor/Vendor _____   Date _____

Description of Need for Backcharge: _____

_____

_____

_____

_____

_____

Applicable Contract/Purchase Order Authority (clauses, sections, etc.)

_____

_____

_____

Backcharge Pricing Method: _____

☐ Lump Sum, or ☐ Estimated Cost: _____

Organization ☐ to perform, ☐ which has performed the backcharged work:

_____

Change Order No. for performance of backcharged work _____

(if applicable)

Instructions to Contractor/Vendor:

The cost of the above-described work ☐ is to be entered and identified as such
on your progress payment estimate
form for the next payment period.

☐ is shown on the attached invoice.
Please honor by _____ .

Prepared by: _____
(contract administrator)

Reviewed by: _____
(construction specialist)

Approved by: _____
(construction manager)

CONTRACTOR/VENDOR ACCEPTANCE: _____

(contractor/vendor)

Any work performed as a result of this backcharge will not relieve the contractor/
vendor from its guarantee or any of the other terms and conditions of the
referenced contract/purchase order.

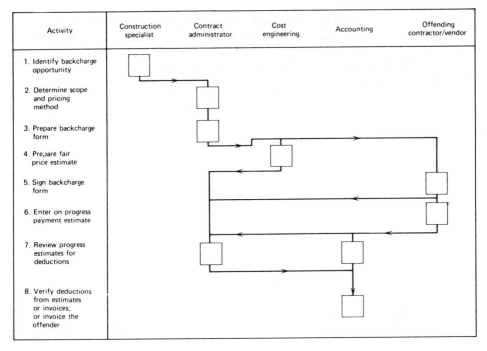

Activity	Construction specialist	Contract administrator	Cost engineering	Accounting	Offending contractor/vendor
1. Identify backcharge opportunity					
2. Determine scope and pricing method					
3. Prepare backcharge form					
4. Prepare fair price estimate					
5. Sign backcharge form					
6. Enter on progress payment estimate					
7. Review progress estimates for deductions					
8. Verify deductions from estimates or invoices; or invoice the offender					

FIGURE 20. Backcharging process.

the vendor replied that the specified motors were shipped earlier by another carrier to the owner's warehouse at an adjacent facility (two miles away), where they had been gathering dust unnoticed.

2. A painting subcontractor requested extra compensation from the project's general contractor for applying prime coats to structural steel that was supposed to be factory primed. The general contractor requested that the owner's contract manager issue a backcharge to the steel supplier for the painting contractor's increased expense. While filling out the backcharge form, the contract manager suddenly realized that the steel in question had not been purchased by the owner, as most had been, but was bought by the general contractor under a recent change order to furnish and erect leave-out steel. Since the offending steel vendor was selected by the general contractor and operating under the general's purchase order, the owner had no reason to invoke a backcharge. The general contractor was told to handle its own problem without involving the owner.

3. A piping contractor began field repairs on coated, insulated steam pipe when it was received at the site with numerous handling tears in the protective coating. Due to cold weather, night work, and unavailability of proper

patching materials and tools, the cost of rework approached the factor cost of the pipe itself. When the vendor was notified, it refused to accept the back-charge, claimed that it would have gladly shipped new sections if it had been apprised of the problem, and added that the warranty on pipe repaired by the owner was void.

CHAPTER 15

CLAIMS

In the context of a construction contract, two principal parties can present claims against each other:

1. The contractor may claim additional time for performance and/or additional compensation from the owner, or some other type of concession (like the lessening of technical requirements or material specifications, for example).
2. The owner may claim relief from the contractor in terms of a reduction in the contract price and/or an acceleration or delay of the contractor's performance.

Of course, many other parties, contractual or otherwise, may present claims to one another or to the owner or contractor. These include subcontractors, consulting firms, or engineering or architectural firms involved. Our discussion will center on claims most commonly encountered during the performance period—from contractor to owner, or from owner to contractor. The same principles of claims defense or pursuit presented in this context also apply to the majority of other claims situations.

A claim is nothing more than a request, or demand, for cost, time or performance compensation, over and above that granted or contemplated, from one contractual party to the other. Claims may be presented in any number of formats, ranging from an informal or even verbal request to a voluminous, highly structured and heavily documented claims package. A common misconception is that claims are litigious in nature—that is, one party is suing the other for damages in a legal sense. This is generally not the case. Although some claims fester to the point where resolution requires a lawsuit or arbitration of some kind, most are resolved long before this occurs.

The vast majority of claims, be they initiated by owner or contractor, are settled through negotiation, adherence to the terms of contract, or some mutually agreeable adjustment in time and cost of performance among the owner and contractor(s). In this era of litigation and threat of litigation, most reasonable owners and contractors realize that resolution without litigation is far more desirable. Both parties usually suffer when claims progress, or are transformed, into lawsuits.

The goal of everyone concerned should be to identify claim situations quickly and resolve them as soon as possible. For owners, the challenge may be put even more simply: Would you rather have a completed project or an enforceable claim in court? Most reasonable owners, and contractors, would choose the former.

CAUSES OF CLAIMS

Like informal change orders, claims originate everywhere. There are an infinite number of causes for claims but almost all have their basis in the allegation that acts, or omissions, by one party to the contract—or, less often, by third parties, acts of God, or others—caused the claiming party damage of some sort. In the highly complex environment of a construction project, given the time and cost pressure on all parties and realizing the myriad of relationships, responsibilities, duties, and interdependencies, one can easily see why claims are as common a feature of the construction landscape as are concrete and structural steel.

Let's first dispense with the case of an owner claiming against a contractor, for it is by far the less common of the two claims situations mentioned earlier. In general, owners file claims against their contractors (and, for that matter, engineering firms or other consultants) for one or more of the following reasons:

1. *Defective Work.* Owners not satisfied with the contractor's product may claim damages that include the cost of repair, replacement, or removal of the defective work. In most cases the work does not meet the contractual specifications or is otherwise not fit for its intended purpose. Occasionally, the goods or services in question do not meet express or implied warranties provided by the contractor or its suppliers.

2. *Delays Caused by the Contractor.* If a contractor has obligated itself to perform the contract work, as a whole or in part, by a specified time, the owner may claim damages when a delay was caused by the contractor, or in other cases, even if the delay was beyond the control of the contractor. Typical damages claimed by owners in this regard are loss of use of the facility, the ripple effect on other contractors, and the increased cost of other delayed work.

3. *As a Claims Defense.* Owners faced with claims represented by contractors may retaliate by pressing a counterclaim. This counterclaim usually attacks, or attempts to discredit, the elements of the original contractor claim—by uncovering overlapping or duplicate cost damages, citing contractor contributions to the claim-provoking conditions, or citing changes or claims clauses in the contract that prohibit or modify the contractor's actions in the event of a dispute, for example.

Another type of claim, though rare, arises from termination or breach of contract. This generally occurs when a contractor fails to complete the work or for some reason leaves the job site. Owners in this situation usually demand to be compensated for the increased cost, over and above that paid to the contractor, for completing the work through other means. Contractors also claim damages when they feel they have been unjustly removed from a project or otherwise prevented from completing their work. Both of these situations arise when the contract, in effect, has been terminated by one party. The focus of our discussion will not be this extreme example, however, for we will concentrate on the more common claims situations—those that arise during the performance period and when continued performance of contractual obligations by both parties is contemplated.

In this regard, most claims found on construction projects are pressed by contractors against the owner for one reason or the other. Their causes are similar, if not the same, as those for informal changes discussed in Chapter 13. These are summarized below:

1. Late or defective owner-furnished information, generally in the form of drawings or specifications
2. Late or defective owner-furnished material or equipment
3. Changes in regulatory requirements, drawings, or specifications
4. Changed or unknown site conditions
5. The ripple effect of collateral work
6. Restrictions in work method, including delay or acceleration of contractor performance
7. Ambiguous contracts or contract interpretations

In each of these situations, the contractor will claim that something has occurred (or failed to occur) that caused it to incur additional cost or spend additional time beyond that called for in the contract, or that could have been reasonably expected at the time of bidding or contract award.

ELEMENTS OF A CONSTRUCTION CLAIM

If any of the claim-provoking circumstances described earlier occur, a conscientious contractor will immediately notify the owner of the causes and effects of each. When this notification is done by pressing a claim, most contractors will request additional time and/or compensation for (1) increased costs of performing the changed work, and (2) "impact costs" on unchanged work. In most cases where a bona fide claims situation has occurred, the contractor has suffered some increased costs (be it in terms of time, money, or both) in each of these categories.

It is fairly easy to identify and, for the owner, accept or defend against direct costs of performing changed work. However, the impact costs—cost on unchanged work—are not as easily defined or priced. Let's deal first with the costs of performing changed work. Some of the most commonly cited costs are for:

Increased labor costs (additional or higher-paid workers)
Additional material or equipment required
Additional supervision, administration, and overhead
Increased time required for performance
Ripout and rework
Decreased productivity or efficiency
Weather effects
Schedule disruption and delays
Demobilization, remobilization
Redundant material handling
Shift-time and overtime premiums
Excessive overtime, leading to production decline
Misallocation of equipment
Loss of material economies
Congestion in the workplace
Inefficient crew sizes or mixes

As for impact costs, all of the preceding may be claimed. The difference is that it is more difficult to determine the nature of the impact and to quantify the increased costs. The question of impact costs can be reduced to a simply put, but difficult to answer, version: What were the increased costs of performing B and C after A was changed? To answer this question, both contractor and owner must determine what it *would* have cost to perform B and C had A *not* changed. This requires a more qualitative analysis and is quite often the most difficult issue concerning impact costs.

The best way to illustrate impact costs is through an example. Suppose the owner somehow delays the contractor's performance and causes it to postpone work it planned to perform during the summer months into the winter season. The work itself remains the same, yet the contractor must incur costs associated with winter work over those anticipated for summer performance. These "impact costs" may include:

1. Cost of protecting the work from the winter weather
2. Inefficiencies in production caused by workers performing in cold weather, with cumbersome protective equipment, on slippery surfaces, and during shorter periods of daylight
3. Costs of heaters and fuel for personnel protection and for actual work performance, such as for heating concrete
4. Increased maintenance costs for equipment
5. Damage to materials and equipment caused by the weather
6. Inability to secure a work force
7. Lost time due to extreme temperature or climatic conditions
8. Increased housing and transportation costs
9. Delay due to Christmas holidays
10. Extension of insurance premiums or bond payments
11. Increases in labor, material, equipment, and overhead costs due to inflation and escalation

THE STRUCTURE OF CONTRACTOR CLAIMS

As mentioned earlier, contractor claims may vary in formality and content. A typical claim, however, generally follows this structure:

1. A description of the contractual terms and conditions, such as scope of work and pricing structure, covering the segment of work in question
2. A factual description of events that have (or have not) taken place, usually presented in chronological order and referencing correspondence, change orders, meetings, and so on
3. The results of the claim-provoking circumstances, usually presented as a narrative description of the increased effort required of the contractor
4. A cost analysis, which may include a detailed listing of increased costs due to the change or a comparison of actual costs and anticipated costs—the difference comprising the claim amount

Remember that a claim is different from a contractor's quotation resulting from a notification or change order. In the strictest sense, they may be the same, considering that the contractor is presenting increased cost information to the owner. However, a quotation occurs *before* the work is done, and a claim is generally presented after, or during, performance of the affected work. Once increased compensation or time for performance has been agreed upon, claims should be transformed into change orders.

CLAIMS ANALYSIS

To judge the merits of a claim and determine what additional compensation, if any, should be allowed, the owner must thoroughly analyze the claim in three stages: (1) a factual analysis (What happened?); (2) a legal or contractual analysis (Is the contractor due relief?); and (3) a cost analysis (How much additional money or time should the contractor be granted?). Factual and legal analyses of claims are easier if you have had proper formation controls, detailed record keeping, structured change control, objective progress and payment determinations, and so on. It is surprising, however, how widely cost analyses may vary given the same factual and legal circumstances. This is the area of claims defense—cost analysis—that contains the highest degree of risk and requires the most owner attention.

There are two distinct methods to quantify the costs of a claim-provoking situation:

1. The total-cost method
2. The incremental-cost method

With the total-cost method, the contractor simply compares the actual cost of performing the work, or work segment, with the anticipated cost (or bid or contract price). The assumption is that *all* increased costs incurred by the contractor are due to the claim-provoking situation. Needless to say, most owners respond negatively to the total-cost approach. The major problem is that the contractor must prove that the work, as changed, was performed as efficiently as possible. This is difficult to do. Although this approach may provide a useful "upper limit" to the claimed cost, it is generally ineffective in pursuing recovery.

The incremental-cost method is recommended over the total-cost method for several reasons. First, it separates increased costs arising from other conditions from those due to the facts of the claim (contractor inefficiency, bad luck, or factors not pertinent to the claim itself). Second, this approach allows costs to be estimated for discrete elements of work under fairly well defined cost parameters. Often with the total-cost philosophy, an unjustified element of increased cost, when included with the claim, obscures or taints the remaining meritable elements, thereby reducing the effectiveness of the claim. And the incremental method focuses on cause and effect in a one-to-one fashion. With the incremental method, contractors relate each additional cost to each factual cause, for example, "Your direction that soil be hand tamped instead of machine rolled caused us the following increased costs." Most important, the incremental cost method allows incremental resolution—easily resolved elements can be separated and handled quickly, while more disputed ones are pending.

DELAY CLAIMS

One of the most common claims is that the owner, architect–engineer, other contractors, or site conditions caused the contractor delay. In most cases, the claim is for additional time and money. Most courts have recognized three distinct types of delays, and their resolution depends on which type is involved. They are:

1. *Excusable Delays.* For these, the contractor is granted a time extension but no additional money or other relief.
2. *Compensable Delays.* Here the contractor is granted not only additional time (if it can demonstrate that it is needed), but additional compensation as well.
3. *Concurrent Delays.* The idea here is that the delay is partially the fault of the owner and partially the fault of the contractor—that the delay periods overlapped, or were *concurrent*. For example, the owner might have been late delivering equipment for the contractor to install, or getting a building permit or zoning authorization so that the contractor could start work. Suppose that this delay postponed start of work from January 1 until July 1 (six-month delay due to owner). Then further suppose that the contractor could not furnish shop drawings or a proper performance bond (or some other requirement) that were due April 1 until July 1 (three-month contractor delay). In other words, the contractor would have been three months late due to its own problems, regardless of whether the owner had been late. The time period from April 1 until July 1 is the overlap—the concurrence. If all this is proven, the contractor would be granted only three months' extra time (and associated costs) for the period January 1 through April 1—the three-month delay due solely to owner problems. When both parties are at fault, the delay is often termed concurrent, and no one is granted relief. Concurrent delays really get complex when delays by one contractor cause delays for other contractors, subs, and so on. Untangling them is a tremendous challenge and usually requires extensive expert analysis, as-built CPM scheduling programs, and the like.

Delay claims almost always lead to requests for time and money. Some common cost items that increase as a result of time (time-sensitive costs) are:

1. Interest
2. Insurance
3. Home office overhead
4. Utilities
5. Rentals

6. Equipment maintenance
7. Supplies
8. Engineering support
9. Contract administration
10. Quality program administration
11. Security
12. Supervision
13. Extension or loss of warranties
14. Demurrage
15. Material storage and protection

Owing to all these, and more, it is easy to see why excusable delays are rare—if time is granted, money is usually granted, too (compensable delay). The most common use of excusable delay is when it is granted in advance— the owner and contractor agree to a postponement for the convenience of one or the other, or both.

OVERHEAD COSTS IN CLAIMS

Can contractors charge owners for overheads simply because of a delay? In other words, if the owner delays the contractor by two months, and besides the direct and impact costs mentioned above the contractor claims a charge for home office overhead—should it be paid? Many people reject this notion out of hand, but it has merit—and it has been upheld in case after case.

What makes up overhead costs? Consider, for example, the salary of the contractor's president, its corporate staff expenses, utility bills on its corporate headquarters, rent, real estate taxes, advertising expenses, and on and on. These are costs that are not charged particularly to any contract, but are accounted for by spreading them over the range of contracts and included, indirectly, in contractor bid prices. If the contract that is delayed stretches from one to two years in duration, these costs go on regardless—they are not proportional to the amount of actual work, or billings, that occur.

Courts have allowed overhead costs in delay situations. The rationale varies, but usually revolves around (1) the contract documents (What do they say about this cost item?); (2) Whether the costs claimed are allowable (Are they included in other contract prices?); (3) what elements are included (president's golf club membership?); and (4) how these corporate costs are allocated over the range of contracts the contractor is active with during the time in question. This final issue gets into fairly esoteric accounting principles that deal with the pool of costs, the burden allocation method used, and the base on which the burden is spread.

Suffice it to say that costs of delay are often much higher than owners realize and should be considered very carefully before delays are approved or allowed to occur.

Another important point surfaces here. There's no excuse for lack of training or awareness on these issues—for claims insensitivity. If you are involved in construction, in any regard, you need to know how time and cost-sensitive your decisions are and be prepared to enjoy or suffer the consequences *before* they are taken. This means that everybody needs to know the fundamentals of claims offense and defense, the types of costs that could be involved, and the criticality of contract management. It also means that cost and schedule systems are vital not only for project control but also for claims protection. The same is true for documentation, contract reporting, and meticulous record keeping. Claims that evolve into lawsuits often rear their ugly heads years after everyone responsible has moved on or forgotten what happened. It just takes getting entangled in one of them to make you a believer in contract management and project control systems.

ARBITRATION TO RESOLVE DISPUTES

Because litigation is such a ponderous, expensive process, most contracts contain an alternative: arbitration. The parties agree in advance to contractual terms detailing when and how arbitration comes into play, who can demand it, and what arbitration organizations and rules will apply. In the United States, those of the American Arbitration Association are by far the most common and most thorough.

Most people cite arbitration's advantages as including (1) low cost, (2) speedier resolution, (3) involves construction or engineering experts rather than attorneys and judges who might know very little about this industry, and perhaps (4) can be done on a more informal, even friendly basis. In general, all of these benefits hold true, but there are drawbacks or other factors that you should consider.

For example, do not assume that the "experts" used are well versed in the technology or practices at hand. Although arbitrators are usually screened and qualified by the association they represent, this can be a very perfunctory assessment of their competency. Then there is the question of a written record. Legal proceedings always result in a printed record, but arbitration need not. And legal judgments are open to appeal—not always so with arbitration.

All things considered, arbitration is a very attract
sure you define all its elements in advance—in the con
the burden of proof and the cost, schedule, and techi
for litigation are not precluded simply because you
nally, learn about arbitration so that you know what y
choose it.

SUGGESTED PROCEDURE FOR HANDLING CLAIMS

As with the change-order process, the contract manager is usually charged with handling claims from their inception through their eventual resolution. And as with several other processes described here, he or she serves more as a coordinator of others than as the authority for the owner. Figure 21 shows the process involved when changes evolve into claims.

The first objective of contract management is to eliminate or reduce the incidence of claim-provoking conditions and claims themselves. When a claim is presented, however, the contract administrator should ensure that it is given proper management visibility, carefully analyzed and documented, and equitably resolved as soon as possible. Few things are as damaging to the morale of owner and contractor personnel alike as bitter, unresolved claims.

All conversations, correspondence, supporting documentation, and the like, pertaining to the claim should be obtained and compiled. A separate file should be maintained for each claim and should include copies of those portions of the contract pertaining to the claim, results of factual, legal, and cost analyses, and other documents that may assist in rejecting or resolving the claim.

Each claim received by the owner should be routed to the contract manager. He or she then opens a claim file and records the claim in a proposal or claim log for tracking purposes. In additional information is required of the contractor, this should be noted in the log and the contractor requested to provide it.

The contract manager then conducts a detailed research effort that includes, among other tasks, interviews with affected owner personnel, review of contract and project files and reports, and compilation of documentation that will be required to analyze the claim. This documentation could include such items as the contract itself, change orders, extra work summary and approval forms, meeting minutes, contractor correspondence, schedules submitted to and by the contractor, project photographs, daily activity or progress reports, time sheets, progress estimates and invoices, and telephone memoranda.

Once the claim file is complete, the contract manager enlists the assistance of project personnel in analyzing the claim and preparing a response to the contractor. These might include the construction manager, design engineers, cost engineers, project schedulers, and the like. Their effort should be structured and coordinated with the goal of an objective analysis and a reasonable response. The contract manager should never agree with the contractor to increased payments or other relief without proper authority, and it is recommended that the project manager, or chief owner representative, conduct all negotiations with the contractor. Again, the contract manager's role is adjunct to the contractual authority of the owner, not necessarily as an enforcer of the contract, although this is permissible if the owner desires.

When a claim is resolved, a change order should be issued to cover the

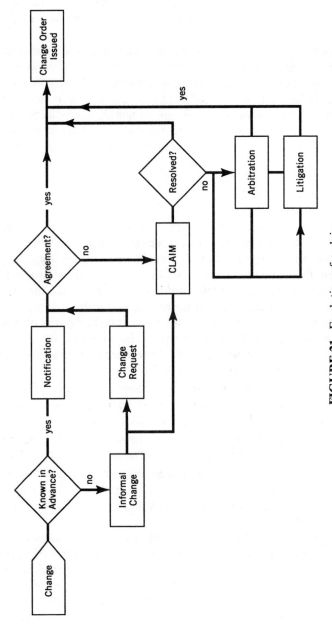

FIGURE 21. Evolution of a claim.

resolution. In this way, all changes to the contract, be they initiated by the owner (a notification and resulting change order) or by the contractor (a claim and resulting change order) are controlled and documented in the same manner.

If resolution is impossible without engaging higher owner management, the contract manager prepares and presents the elements of the claim and its analyses and assists them in proceeding as required. Although the tendency is strong, the contract manager should not become emotionally involved in the dispute, but instead accept the challenge to handle each claim on its own merits, to involve and coordinate the efforts of affected or responsible project personnel, and to structure and document the process.

REPRESENTATIVE CASES

1. The owner of an industrial complex under construction had recently sent its chief mechanical engineer to a three-day seminar on weld inspection techniques. When he returned, full of enthusiasm for radiographic inspection to detect weld flaws, he revised the specifications for buried process steam-piping just after the contractor had commenced work. Since the inspection was to be done by an independent agency hired by the owner, no change order was issued the piping contractor. After several months of work, the piping contractor presented the owner with a claim for $195,000 in additional compensation due to inefficiencies and interference caused by the increased inspection process. Briefly, the claim cited:

The original specs called only for visual weld inspection. Periodic x-ray inspections caused pipe welding to cease in open trenches, as welders did not want to be exposed to the radiation closeby.

Many more welds were rejected, causing rework and schedule extension into the winter season and resulting in inefficiency in weld production.

Additional welds necessitated by rejects required four additional crews. Since qualified welders were scarce in the project location, it paid extra to "import" welders and train or recertify others.

2. A unit-price contract for a fossil-fired power plant was awarded to an electrical contractor. Drawing changes concerning underground cable ducts and cable runs were issued in a flurry once cable pulling began in the plant. The changes had the overall impact of increasing the length of cable pulled by only 10% over the original estimate, and the contractor was paid by the agreed unit prices for these additions. It submitted a claim, however, for compensation over and above this amount to account for:

Inefficiencies in operation when cable was measured, cut, pulled, and then removed and discarded as revisions to drawings changed cable size and routing

Inefficiencies in cable purchasing and cutting, because the contractor could not plan the usage of cable to maximize usable lengths cut from standard purchased spools

Demobilization, standby, and remobilization of work crews from one area of the plant to others as revised drawings became available

Increased labor cost caused by drawing changes that postponed much of the work into a new labor-contract year, with higher labor rates in effect

3. A lump-sum contract for installation of the turbine generator for a nuclear power plant was awarded to a mechanical contractor (contractor A). The equipment was to be delivered by barge two weeks after contractor A mobilized at the site. A heavy freeze caused ice blockage on the river adjacent to the plant, with a two-month delay in receipt of the owner's turbine generator equipment. To make up for lost time, the project manager ordered the general contractor (contractor B) to begin installation of 110-inch-diameter circulating water pipe from the turbine building to the cooling tower. When the turbine generator finally arrived, contractor A could not transfer the heavy components from the barge unloading area to their destination on the turbine pedestal in the turbine building because two 20-foot-deep pipe trenches filled with partially installed circulating water pipe blocked its access. Contractor A filed a claim for increased compensation caused by:

Standby of crews and equipment for the two months of ice delay and additional two months of pipe trench delay

Temporary storage of the turbine generator on site

Acceleration of work once the trenches were covered to compensate for lost time

Loss of profit caused by being unable to use its workers and equipment on other jobs

CHAPTER 16

SHORT-FORM CONTRACTING

Not all construction requirements can be seen from the perspective of a home office several months or years before they occur. As a consequence, some flexibility should be given to site management in order to react to changed or unforeseen events requiring the contracting process. Many owners implement short-form contract documents and procedures for this purpose. Two types of contracting instruments are generally involved: (1) field-issued short-form contracts, and (2) field change orders to home office–issued contracts.

Without short-form contracting instruments, unacceptable contracting methods may be implemented in their place. These include the use of purchase orders, verbal agreements, telephone orders, or the often-abused handshake technique. To ensure that all job-site labor is covered by an express, written contract containing protection for all parties, avoid these methods in all cases. But when the alternative is a lengthy formation process or unnecessarily complicated documents, this is impossible. The need exists for an alternative method of contracting that is rapidly achieved and easily administered and at the same time provides formation and administration controls. The answer is short-form contracting.

SHORT-FORM CONTRACTS

The big problem with home office–issued contracts is that their formation takes a long time. It can take several months to qualify bidders, create the RFP, prepare bids, and receive and evaluate proposals before award. In addition, "long-form" contract documents often contain commercial terms that are unnecessary for short-term, low-cost scopes of work. Most owners use long-form contracts for all foreseeable work and for scopes of work that are medium to large in amount, risk, and cost.

Short-form contracts are designed to apply to other conditions. They are used for short-duration, low-cost work that cannot be planned and priced well in advance. Unforeseen events such as emergencies and planning errors fall into this category. Where the typical long-form contract can be hundred of pages in length, a short-form version may be written on a dozen or less. Bidder qualification and other formation processes are performed in an abbreviated fashion, and a contract agreement may be formed in a few days or less, rather than over a period of several months. Common applications for short-form contracts include:

1. *Emergency Work.* Repairs caused by fires, storm damage, or safety measures fall into this category.
2. *Housekeeping Services.* These include general cleanup and maintenance services performed during construction. Examples are snow removal, trash collection and disposal, janitorial services, guard services, and sanitary services.
3. *As Relief for Large Contractors.* Should an existing contractor be unable or unwilling to perform additional work, smaller, local contractors may be hired to perform the work.
4. *As Competition for Change Orders to Existing Contractors.* In some cases cost quotations from existing contractors to perform additional work may appear to be noncompetitive or excessively high. If the work may be performed by other, locally available contractors, they should also be asked to quote. This provides some means of protection from an existing contractor that feels it holds the owner captive by its presence at the job site. Or from a sub who feels the same about the general contractor.
5. *When Existing Contractors Have Been Performing Unsatisfactorily.* If additional work is required and the owner is dissatisfied with the available sources at the site, short-form contracts may be used to hire other contractors capable of better performance.
6. *To Perform Backcharge Work.* At times there is no site contractor available or capable of performing backcharged work.

Although their use must be carefully controlled, short-form contracts have significant advantages. They can be rapidly formed. The cost of formation is often insignificant when compared to the cost of forming long-form contracts. They are ideal for small scopes of work, where the formation of complex contractual documents exceeds or is out of proportion to the work to be performed. They are excellent when used for rental of construction equipment. They can also be used to relieve large contractors from the requirements of standing down or remobilization to complete minor sections of work.

PROCEDURE FOR SHORT-FORM CONTRACTS

Short-form contracts should be formed and administered with the essential elements of control required for long-form versions. Competitive bidding by qualified contractors should be encouraged, though the degree of qualification and the number of available bidders are generally reduced. Standard or reference contract documents should also be prepared, each containing appropriate clauses. And administration documentation should be maintained throughout the performance period. The following suggestions should be considered when short forms are used:

1. To the extent practical, short-form documents should parallel the content and format of long-form contracts.
2. Competitive bidding should be encouraged if possible.
3. Existing contracts, or those planned for the future, should be reviewed before the need for short-form contracting is established.
4. The alternative of adding work to an existing contract (change order) should not be overlooked.
5. Close coordination between long-form contracts, typically formed in the home office, and short-form contracts, typically formed at the site location, should be maintained to avoid duplication of effort, redundant scopes of work, or loss of scope entirely.
6. Administration of short-form contracts should be identical to other contracts.
7. Limitations on short-form contracts should be established and enforced. Most owners stipulate certain cost, time, or scope limitations. For example, any contract priced at over $50,000 may require long-form documents and the standard formation processes described in Part 2. A common tendency, however, is for these "thresholds" to be exceeded—for short-form contracting to be used for larger cost items. Owners should make sure that the thresholds bear a resemblance to actual project conditions and needs. If they are unreasonably low, they should be raised. If they are sufficiently high, they should be enforced. And actual contracts should be reviewed to ensure that a series of short-form contracts is not being used if the work could be consolidated into one larger, long-form version.

PROTECTION VERSUS BREVITY

For all the benefits of short-form contracts, there is the risk that protection is lost by virtue of their condensed form. Needless to say, if any contract clauses were unnecessary, they would also be eliminated from long-form documents. So why are short-form documents "shorter" than long-form versions? The answer lies with the fact that long-form documents are used for special conditions not anticipated for the work that is done under short-form applications—such conditions as:

1. *Progress Payments.* With short-duration work, provisions covering the measurement and payment for progress may not be required; final payment may be all that is involved.
2. *Changed Work.* Since the nature of short-form contracting is generally to cover changed work, changes to short-form contracts may not be required. Should additional work be necessary, it may be covered by a change order to another contractor or through the use of another short-form contract.
3. *Contractor Submittals.* These may not be required.
4. *Equipment Warranties, Titles, Spare Parts, Tools, and so on.* Most short-form contracts are service related, without the need for complex clauses covering equipment and materials.
5. *Closeout Documentation.* Most likely, the short-form contractor is not supplying as-built drawings, releases of lien, or other documents.
6. *Technical Specifications and Drawings.* Short-form contracts are generally reserved for low-technology work, and there may be no need for technical specs or drawings to define scope. Where used, the specifications may entail just a few sentences or paragraphs.
7. *Required Contractor Controls.* Sophisticated cost and schedule control systems, planning of work, progress reporting, and cost records may not be required.
8. *Retention.* Because of duration and scope of work, this may not be employed.

These are just some reasons why short-form contracts can be streamlined. Of course, significant danger lies in using such contracts when the cost, scope of work, or contract duration exceeds that envisioned by the abbreviated documents. Because of their coverage limitations, short-form contracts should never be used when they do not contain provisions for anticipated or potential needs of the work in question.

SHORT-FORM CHANGE ORDERS

Owners often use two distinct versions of change order documents for the same reasons that they use two versions of contract documents. Short-form change orders are used for field-originated changes, emergency work, or minor alterations to the contractor's scope. Long-form change orders are used for changes originating in the home office, such as design revisions, for changes that are planned well in advance of incorporation, and for changes requiring significant time to consider, evaluate, and approve. Short-form change orders are then used for the remaining change situations—those that are unplanned, result from site-specific conditions (bad weather, interferences, and so on), and cover immediate needs caused by unforeseen events.

Just as with short-form contracts, short-form change orders are designed to prevent unacceptable contracting practices that may be used to respond to unforeseen and pressing events. They should be identified, evaluated, and approved with the same degree of control as their home office–issued counterparts. Give particular attention to the possibility of duplication between home office change orders, field office change orders, home office contracts and field office (short-form) contracts.

SHORT-FORM CONTRACTING CAUTIONS

For all of their advantages, both short-form contracts and change orders contain special dangers and are often misused for the sake of expediency at the expense of control. They should be carefully designed and limited to specific applications, and their use should be continually monitored. Because of their special nature and the environments under which they are used, certain additional cautions apply:

1. Because short-form contracts are often used with smaller, local contractors, pay particular attention to insurance coverage, union affiliations, if applicable, and bonding capacity, where required.
2. Establish procedures to eliminate possible duplication of effort and payment between short- and long-form applications.
3. Review short-form documents prior to each application to ensure that critical contractual protection is not lost by the abbreviated nature of the documents.
4. Avoid the tendency to price short-form contracts or change orders in a cost-reimbursable manner when other fixed-price methods may be used.
5. Monitor adherence to time, cost, or scope limitations.

REPRESENTATIVE CASES

1. On a project to build an addition to an automobile assembly plant, a fire broke out over a weekend and gutted the facility's existing warehouse. Since the warehouse was needed for storage of equipment and material to be used in constructing the addition, it was necessary to demolish the gutted portions and make temporary repairs as soon as possible. A short-form contract was issued to a local general contractor to perform this work.

2. Due to expansion of site office staff, the construction manager for a pulp mill project decided that additional temporary parking areas adjacent to the

office trailers were needed. Upon checking the local Yellow Pages, the contract manager found a grading and paving contractor willing to do the work for $12,000. The response to a notification to the site civil contractor was a quote of $26,000 if this work was to be made a change order. The local paving and grading contractor was awarded the work by a short-form contract for $12,000, and no change orders were issued to the site civil contractor.

3. Sheet pile driving at a process plant's intake water pumping station was supposed to take place during the dry season. Heavy rains and the failure of a small upstream dam flooded the area after the piles were driven but before final inspection of pile splices could be made. Underwater divers were required to make this inspection, and the company providing these specialty services was engaged through a short-form contract.

4. A short oil pipeline project required the crossing of an unguarded railroad line by material delivery trucks. For safety reasons, the owner decided to secure the crossing with a flag-guard during working hours. A guard service was hired to provide this worker under a short-form contract.

5. The owner of a nuclear power plant under construction decided to hire certain construction cranes for use in the area of the reactor building rather than to require several contractors to provide their own. The owner scheduled their use and prorated their cost among the using contractors. The fully operated cranes were rented under a field-issued equipment rental contract.

6. The painting contractor for a large industrial plant project had completed all of his painting except for a small masonry auxilliary building that had been delayed due to a shortage of concrete blocks in the area. Rather than keeping the painting contractor on standby at $500 per day, the owner closed out its contract with a credit of $6,000 for not painting the building in question. Four months later, the building was constructed, and a local contractor was awarded an $8,500 short-form contract to paint it.

CHAPTER 17

CONTRACT CLOSEOUT

Construction contracts should be formally terminated once performance has been completed and final payment has been made. Because final payment and the release of remaining retention constitute release of a vital compliance control (the payment of money), a structured process leading to this release is recommended. This process, called contract closeout, involves final acceptance of the contractor's work, receipt of acceptable documentation required by the contractual terms, and an evaluation of the contractor's performance. Though it is performed with the assistance of many people, the contract manager should coordinate and control contract closeout. This chapter describes the closeout process, discusses the responsibilities of those typically involved, and presents a recommended procedure for its conduct.

DOCUMENTATION INVOLVED

Depending on the contractor's scope of work, the owner's objectives, and the specifics of the contract in question, closeout generally involves the receipt of documentation from the contractor. To ensure that it is submitted and is acceptable, final payment and release of retention (if used) should be withheld until it is received and approved. Commonly required documentation includes:

1. *Releases of Liens from the Contractor and Its Suppliers and Subcontractors.* These are formal documents signed by those who have performed work or provided materials or services in conjunction with the contract. They are, in effect, promises that no liens have been filed or will be filed. Their format and content should follow applicable legal requirements.

Some owners insert sample release forms in the contract documents at the time of award. These releases constitute strong protection for two reasons: (1) they may be enforceable in court when a subsequent lien is filed, and (2) when made a prerequisite to final payment, they force the contractor (and its suppliers and subcontractors) to examine their books in order to detect any lack of payment or failure to receive payment. Should either have occurred, it is much better to present these problems at the time of closeout than afterward. They can then be rectified before conditions leading to a lien situation arise and interfere or prevent the owner's occupation or use of the completed facility.

2. *Titles to Major Equipment Incorporated in the Facility.* Ownership of certain pieces of equipment and machinery may need to be transferred to the owner through clear titles. These should be received and verified before final payment is made.

3. *Warranty Documentation.* Operating equipment is generally sold with a specific warranty, usually in effect for one year. The warranty documentation for this equipment should pass to the owner in accordance with customary practices or specific contractual terms.

4. *As-Built Drawings.* Most major facilities are not constructed in exact accordance with the drawings issued prior to or during the construction effort. This is not to suggest that drawings are not followed, for they should be. There are many cases, however, where the drawings do not show exact locations, positions, relationships, and sizes. Instead, they give general locations, suggested positions, and general dimensions—allowing the constructor tolerances and flexibility in installation or erection. In other cases, changes to the work, for the benefit of the owner or to simplify construction itself, are made as the construction process is carried out. When these changes depart from the drawings, actual as-constructed positions or locations should be shown on a separate set of drawings called "as-builts."

A common example of as-built conditions that differ from design or suggested conditions involves small diameter piping that is routed according to field conditions at the time of its installation. Most design drawings leave this routing to the contractor. Naturally, the owner should check the routing to make sure that there are no interferences with other objects and that the piping is supported soundly, achieves its purpose, and is installed correctly. But the details of its installation are lost unless described by the as-builts. Other examples include minor adjustments for field fit up of equipment, conduit, cable tray, ductwork, and the like. As-built drawings are important. They often represent the only accurate record of the constructed facility and are often needed for operations, maintenance, and repair throughout the facility's life. Should renovations or retrofits be needed years later, the as-builts allow these to be designed and constructed with full knowledge of what is "out there" (or "under there"). Much of the completed work may be buried underground, enclosed, or otherwise inaccessible. Millions of dollars have been spent locating and measuring such items when representative as-builts were not available. Three specific recommendations are given for dealing with as-built drawings: (1) They should be thoroughly checked and verified before approval. (2) Should the cost of their preparation be included in the contract price, make sure that the owner does not incur this cost a second or third time. This can happen when the engineering firm or others prepares as-builts instead of (or concurrently with) the contractor. In addition, (3) approved as-builts should be secured and protected once received, to be easily located and understood years later.

5. *Inspection and Acceptance Records.* These include test reports, calibration data, and inspection results. Although initially required to ensure that the work was performed correctly and that the finished products are acceptable, they also provide valuable information that may be needed at a future date.

6. *Operating and Maintenance Manuals.* The need to operate and maintain the finished facility is often overlooked in the rush to complete construction and closeout the contract(s). Owners should receive appropriate documents needed for both purposes. Training material relating to the actual operation of completed systems, processes, or equipment should also be received where needed.

Other Typical Deliveries

In addition to closeout documentation, other items may be required before the contractor is released of its obligation. Examples include:

1. *Spare Parts.* These are especially important for specialized equipment and equipment that is custom made for the project. Critical equipment and equipment prone to failure should be furnished with some spare parts; the owner should identify these and make their receipt part of the contract requirements. This is especially true for equipment having long lead times to replace or repair and equipment supplied by only a few sources.
2. *Special Tools.* Many constructed or installed items require special tools not normally maintained by the owner. These should be identified, required by the contract, and received prior to final payment.
3. *Consumable Supplies.* This category might include special fuels, lubricants, gaskets, seals, valve packing, electrical parts (light bulbs), cleaning solvents, and liquid or gaseous charges.

Needless to say, it is difficult or impossible to obtain spare parts, special tools, or consumable supplies unless they are specified by the original contract documents. They should be considered not only when the contractor is about to depart but also as part of the contract formation process. If continuing service agreements are made part of the construction contract, the prices and terms of both services and materials should be covered as well.

SUGGESTED CLOSEOUT PROCEDURE

Owner control over the construction process typically declines toward the end of the project. This is generally because of the schedule pressure commonly encountered, the reduction of supervisory personnel as the work winds down, and a commonly felt urgency to finish the work and move on to newer challenges. Unfortunately, however, this is the time when control is perhaps needed most. In order to manage the closeout process, owners should design and implement structured procedures well in advance of this period.

Exhibit 21 depicts a general closeout checklist that is designed to ensure that all required items have been received, all work has been accepted, and no commercial transactions are incomplete. It is completed by owner personnel and representatives responsible for verifying specific elements of the contractor's performance—for example, the resident engineer checks the as-builts, the contract manager checks such items as backcharges and unresolved change orders, and construction accounting reviews the final invoice. The completed checklist should be a prerequisite to final payment, along with the final progress payment estimate form submitted by the contractor.

Postperformance Evaluation

As part of the closeout process, consider a thorough and objective evaluation of the contractor's performance by those who have dealt with the contractor in many different areas, including the construction specialist, contract manager, and construction manager. Completed evaluation forms (see Exhibit 22) are returned to the owner's bidder qualification files for use when reviewing the contractor for future work. Make sure that these forms reflect the entire scope of a contractor's performance, that they call for an objective evaluation, and that any problem areas are thoroughly explained. Completed evaluations should not be transmitted to the contractor but should be used only for owner information.

It is also suggested that field monitoring and administration personnel indicate improvements they recommend for future contracts of the same type. For example, if the contract in question lacked commercial controls, contained obscure or conflicting requirements, or was priced in a manner that made determination of progress difficult, these should be noted. In this way, the contractor evaluation becomes an indicator of not only the contractor's performance but also of the contract itself. This will allow improvements in future contracting efforts. Some companies use a separate form to evaluate the contract itself. An example is shown as Exhibit 23.

EXHIBIT 21.   Contract Closeout Checklist

Contract Closeout Checklist

_____
Owner

_____
Project

Contract No. _____
Contract Work _____
Contractor _____

Check Item	Yes	No	N/A	Initial
1. Engineer final inspection	____	____	____	____
2. Corrections complete	____	____	____	____
3. As-built drawings received	____	____	____	____
4. As-built drawings satisfactory	____	____	____	____
5. Operating manuals received	____	____	____	____
6. Warranty statements received	____	____	____	____
7. QA/QC documentation complete	____	____	____	____
8. Backcharges satisfied	____	____	____	____
9. Release from contractor	____	____	____	____
10. Final invoice approved	____	____	____	____

Route:	Signature	Date
1. Contract administrator	_____	_____
2. Cost engineer	_____	_____
3. Resident engineer	_____	_____
4. Construction accounting	_____	_____
5. Construction manager	_____	_____

Termination Notification

Once the work has been performed and accepted, the closeout checklist completed, and an evaluation made, the owner should send the contractor a formal letter of termination. This letter informs the contractor that the contractual association is officially ended; it may be sent to others who have participated in the relationship as well. Formal notification is recommended for a number of reasons. It officially notifies the contractor that charges under

EXHIBIT 22.   Contractor Evaluation

CONTRACTOR EVALUATION

Date _____

Contractor _____   Project _____

Address _____   Contract No. _____

Title _____

Contractor's Personnel: Report Period _____ To _____

Project Manager _____   Type of Contract _____

Project Engineer _____   Contract Amount: Original $ _____

Field Superintendent _____   Change Orders: No. _____ $ _____

Scope of Work Performed _____   Final Contract Amount: $ _____

Instructions: Contract Manager in cooperation with the Construction Specialist prepares the evaluation for review and approval by the Construction Manager.

Item	Satisfactory	Unsatisfactory
1. Supervision	_____	_____
2. Planning	_____	_____
3. Support from home office	_____	_____
4. Drawings	_____	_____
5. Material delivery	_____	_____
6. Construction equipment	_____	_____
7. Labor	_____	_____
8. Progress payment preparation	_____	_____
9. Administration of extra work	_____	_____
10. Cooperation	_____	_____
Composite rating	_____	_____

Explain unsatisfactory items (use item number):

Was an undue amount of time spent administering this contract?

What improvements could have been made in this contract?

Would you recommend awarding future contracts to this firm?
Additional comments:

Prepared By _____   Approved _____
Contract Manager   Construction Manager

_____
Construction Specialist

Exhibit 23 Contract Evaluation Form

Project _____ Contract No. _____ Contract _____
Contractor _____ Pricing type _____
Start date _____ Completion date _____ Final cost _____

Evaluation of Contract Documents Used

	Satisfactory	Unsatisfactory*	Remarks
Invitation to bid	_____	_____	_____
Instructions to bidders	_____	_____	_____
Proposal form(s)	_____	_____	_____
Agreement	_____	_____	_____
General conditions	_____	_____	_____
Special conditions	_____	_____	_____
Technical specifications	_____	_____	_____
Drawings	_____	_____	_____
Other	_____	_____	_____

* Attach Contract Document Change Request for all cases where
Unsatisfactory rating is used. CDCR No. _____ attached.

Additional remarks about the
documents: _____
_____
_____
_____

List submittals required but not
needed: _____
_____

List submittals needed but not
required: _____
_____

Problems encountered during contract administration

Problem	Yes	No	Description or Reason
Progress payments	___	___	_____
Changed work	___	___	_____
Backcharging	___	___	_____
Documents and records	___	___	_____
Retention	___	___	_____
Inspection	___	___	_____
Letter of intent	___	___	_____
Pricing method(s)	___	___	_____
Other: _____	___	___	_____
_____	___	___	_____
_____	___	___	_____

Additional remarks: _____
_____
_____
_____
_____

Prepared by: _____ _____ Reviewed by: _____ _____
      (name)   (date)   (name)   (date)

the contract will no longer be accepted—and that no additional work should therefore be performed. Copies should be sent to owner accounting or accounts payable groups for this reason. If special billing or account numbers have been used, these should also be officially terminated. This prevents unwarranted charging of time or other costs. If unresolved claims or change orders still exist, the terminationletter should identify these and describe plans for their resolution. The amount of and reasons for any remaining retention or withholding of payments should also be identified.

Contract closeout signals the end of the contractual relationship between owner and contractor and is the final step in the commercial administration of the contract. When performed under controlled conditions and formally documented, closeout ensures that both parties have performed according to the agreement and that both have received the benefits of the relationship. The owner benefits because it receives the necessary documentation and deliverables required to complete the construction process and operate and maintain the constructed facility. The owner also captures valuable information regarding the contractor's performance and the contract documents used. Both should improve future contracting efforts. The contractor benefits by rapid resolution of remaining items that may impede final payment. Third parties are notified of the termination in order to prevent unwarranted charges or the performance of work outside the contractual agreement. And the relationship that began under structured and formal conditions (the signing of a contract) is dissolved with the same degree of control for the benefit of all concerned.

REPRESENTATIVE CASES

1. Before approving the final invoice and releasing retention for a large piping and mechanical contract, the field accountant conducted several routine checks. He called the field engineer and was told that a roll of as-built drawings had been received. The plant operations department told him a shipment of spare parts and special tools for the installed equipment had been received at the plant warehouse, and the contract administrator informed him that a release of liens form from the contractor was "in the mail." The field accountant vouched the invoice, and in two days the contractor picked up the final check and departed. One week later, the project manager was on the phone to the accountant demanding to know why final payment was made. It seems the as-built drawings were no more than original contract drawings, each stamped "As-built" by the contractor. No modifications due to change orders, field routing of small bore pipe, or changes in underground pipe routing and elevations due to interferences were shown on the drawings. In addition, liens had just been filed in the county courthouse by an unpaid pipe supplier and an excavation subcontractor. Finally, the boxes of spare parts and special tools received by the warehouse were opened. They contained a

crescent wrench, a can of 3-in-1 oil, and a book of matches advertising "Learn to be a mechanic for fun and profit."

2. Mr. X, civil superintendent for the construction of a chemical processing plant, believed that he should be the one to file a contractor evaluation report on the site civil contractor—after all, X had been in charge of the contract. X had no problem with contractor A—they had worked well together over the past two years. Contractor A had remembered X's birthday with a case of premium bourbon, and each Christmas and Thanksgiving his automobile trunk was certain to contain a huge turkey courtesy of contractor A. Besides, X and contractor A's superintendent went way back together, and X had been given an open job offer by contractor A any time he tired of his present position.

X gave contractor A outstanding ratings for its performance and sent copies of the evaluation to each division of the owner's operations. Had he checked with others on the project staff, he would have learned that:

Contractor A was dependably belligerent and demanding when negotiating change orders.

Contractor A, as a habit, was five months late in paying subcontractors. These subs constantly complained to the owner and threatened to cease work many times.

Contractor A's invoices were never in agreement with progress estimates and usually contained clerical errors, overbillings, and duplicate markups for expenses.

Contractor A refused to remove two foremen caught selling drugs on the job site.

Contractor A's shop drawings were of poor quality and consistently late, causing constant rescheduling of other contractors' work.

Contractor A's general liability insurance had been canceled near the end of the job due to repeated accidents on all its projects, as well as countless violations of safety regulations.

Contractor A's construction equipment was continually under repair or not operating due to lack of preventive maintenance and qualified operators.

PART 4

CONTRACT MONITORING

Proper contract planning, the formation of contracts containing compliance controls, and effective contract administration can be implemented at the very beginning of a project—but by which means does management assure itself of *continuing* control, identify problems that may jeopardize project success, or take corrective action when those problems are identified?

Contract monitoring is a term that applies to those systems and activities designed to answer this question. It entails a formal program of contract auditing and reporting. This part will address those techniques that, along with those already presented, allow management to maintain visibility over contracting efforts. It contains discussion of:

1. *Contract Reporting.* Several reporting requirements are identified, and suggested reporting formats, frequency, and information-gathering challenges are presented.
2. *Contract Auditing.* This chapter discusses the role of internal and independent auditors and presents distinctions between the need for and conduct of operational as well as cost audits. It also gives suggestions for audit techniques targeted on risk and proven to work.

CHAPTER 18

CONTRACT REPORTING

Owners need key information to determine the status, performance, and expected results of critical project activities. A multitude of reports are used for this purpose. Among these are procurement reports, cash-flow projections, schedule progress reports, cost forecasts, labor usage reports, productivity reports, and the like. These fall into the general category of "project reports."

Contract reporting is a special subset of project reporting. It deals exclusively with factors that indicate the commercial success, or failure, of the contracting process. The relationship between contract reports and other project reporting efforts is sometimes confusing. Many times contract-specific reporting systems overlap with other project reports and/or conflict with information presented by them. In the design and use of contract reports, a major challenge is to identify key performance indicators over the contracting process, present them to management in a way that allows isolation of problem areas for corrective action, and to ensure that this information is reconciled or integrated with other data used for the project.

PURPOSE OF CONTRACT REPORTING

For most owners, a large portion of the cost of a new facility is committed and paid for contractually. This means direct monitoring of contractual cost data is essential for control not only of contract costs in themselves but also, by controlling contract cost, of the cost of the entire project. Contractual cost information, therefore, is a major element of contract reporting. Project schedule performance is also directly related to the timely performance of contractors and their subcontractors. Because technical performance—the quality of a contractor's work—is the subject of other technical monitoring and reporting efforts, contract reporting should focus on the two remaining elements of risk: cost and schedule.

Reporting systems always take a great deal of effort and result in some added cost. Any contract reporting system should be designed and implemented so that the cost of gaining contractual information does not outweigh the benefit of having it. For this to occur, the reporting objectives described in the following paragraphs must be met.

1. *Status and performance information should be presented in a format that allows analysis.* Nothing is more frustrating and less valuable than information that does not facilitate problem analysis and subsequent corrective action. Well-designed contract reports should contain easily comparable data regarding planned (or budgeted), actual, and forecast performance. In the case of costs, contract reports should show *estimated contract prices, committed prices* (i.e., the contract price once it is awarded), *actual contract prices* (paid to date, and as amended by change orders, backcharges, credits to the owner, etc.), and *estimates of final price* upon completion. These data allow the owner to determine what was expected, what was achieved through formation, what has happened so far, and what is expected to happen in the future. Significant discrepancies among these should point out areas for further investigation, for they indicate that critical problems have occurred or are about to occur.

The most common problem hampering this type of analysis is caused by differences in the structure or format of cost data. Every effort should be made to ensure that contract prices are arranged in the same format throughout the project life, from the structure of prices requested in the RFP to the "line items" on progress payments and the entries on contractor invoices. Otherwise, meaningful comparisons between planned, actual, and forecast amounts are difficult or impossible. Because contract reports typically use the same cost data as other project cost reports, or are even used as input to these reports, they should be structured to allow relatively easy comparison between the two.

2. *Reports should allow monitoring of contractor progress.* Many progress reports depict owner activities but do not give a complete picture of contractor progress. This is particularly true with regard to schedule and cost information. For example, a report showing that of a total contract price of $1 million the owner has paid $500,000 to date indicates the owner has spent one-half of the contract amount. This tells nothing of the contractor's progress. It could be that the contractor has completed one-third of required work, even though it has been paid for half. The use of discrete, objective progress payment techniques helps solve this type of problem, but allowances should also be made for the impacts of escalation, credits to the owner, backcharges, change orders, and work that is partially completed yet not paid for—not to mention the time lag between a contractor's achievement of

progress, acceptance by the owner, and eventual payment for that progress (the invoicing cycle). To be meaningful, therefore, comparisons of planned with actual performance should be precise and current. Instead of comparing current payments with total contract price, then, current performance should be compared with planned performance for the same period.

3. *Contract reports should focus on owner performance as well.* As mentioned earlier, compliance with contractual terms is required of both parties. The most critical, and often overlooked, element of owner performance relates to the timing of owner activities. For this reason, reports should compare scheduled owner performance, such as review of design drawings, review and approval of contractor submittals, and completion of acceptance testing, with actual owner performance. Significant delays between planned, or committed, and actual owner performance could lead to (1) delay in contractor performance, (2) increased cost through informal change orders, or (3) contractor claims. Not only should contractual obligations be monitored, but so should the timing of contract formation activities. Owner management should review the formation cycle by comparing planned and actual dates for release of the RFP, bid evaluation, and contract award in particular. An example contract formation schedule report focusing on these indicators is shown later in this chapter as Exhibit 24.

4. *Contract reports should allow "management by exception."* There is a sea of data that could be directed toward management at any given time. Much of it, it is hoped, demonstrates that things are going according to plan or that no corrective action is required. Many owners, however, feel as though they are "drowning in data but starving for information." *Data* give management the symptoms of contractual performance, and *information* allows them to determine the extent and expected outcome of problems—How bad is it, and will it get better or worse?—isolate the causes, apply corrective action, and forecast the ultimate impact on the project. The difference between data and information can be demonstrated through an example. Suppose that a $1 million contract is in effect for lump-sum work, that $900,000 has been charged the owner, and that approximately 50% of the work has been accomplished. At first glance, something seems wrong, but any number of conclusions may be drawn from these data. Some of them are:

The work is proceeding according to plan. Mobilization costs of $300,000 were paid, as stipulated in the contract, as soon as the contractor arrived at the site. Changed work in the amount of $200,000 has been ordered by the owner, and one-half of this has been completed ($100,000). Half of the original contract work has been completed ($50,000). Although the amount of mobilization costs may be subject to criticism, nothing indicates deviation from cost or schedule plans.

The work is proceeding according to plan. The contractor is actually 90% complete and has been paid for that amount. The 50% completion figure was calculated two days prior to the arrival of a large, contractor-furnished pressure vessel, for which it was paid $400,000.

The contractor's performance is unknown. The discrepancy between cost paid and progress to date is due to a difference between owner methods of (1) measuring progress in order to pay the contractor, and (2) measuring progress in order to determine project status. These two determinations were made by different owner personnel using different progress estimating criteria.

Errors exist in either the measurement of progress, contractor payments, or both.

The contractor has performed $400,000 worth of emergency changed work that has been authorized by site management but has yet to be reported to higher owner management. This amount has not been entered as a contractual commitment.

Because of a claim settlement caused by owner-ordered acceleration, the contractor has been awarded an additional $400,000 for overtime, extra crews, increased material costs, and labor inefficiency. This increase has not been reflected on a subsequent change order.

A major design change was made that caused the contractor to rip out installed material ($100,000) and reinstall it in another location ($300,000).

The list of possible reasons for such a discrepancy could go on and on. It would serve only to point out the need for "problem traceability" within a contract reporting system, as well as to demonstrate the care that must be exercised when planning and collecting cost and performance data.

5. *Information should be timely.* Except from a "lessons-learned" standpoint, information that is available after the fact is without value. It does little good to be apprised of problems once their impact has been made. Of course, it is helpful to know why things went wrong, but it is much more important to be able to know that things are going wrong *when* it is happening. And management should also know, to the extent possible, that things may go wrong *before* they actually do. In order for this to happen, contract reports should be issued frequently and reflect timely information. This "early warning" allows everyone to focus on problem areas before they get out of hand and to direct corrective action toward them as soon as possible.

CONTRACT REPORTING HIERARCHY

Reporting that allows management by exception, data traceability, and timely response to actual or potential problems, requires a structured system such as that depicted in Figure 22. The documents presented there show a recommended arrangement of (1) source documents; (2) contractual tracking and compliance documents helpful to the contract manager; (3) detailed contract reports covering schedule, claims, change orders, and backcharges;

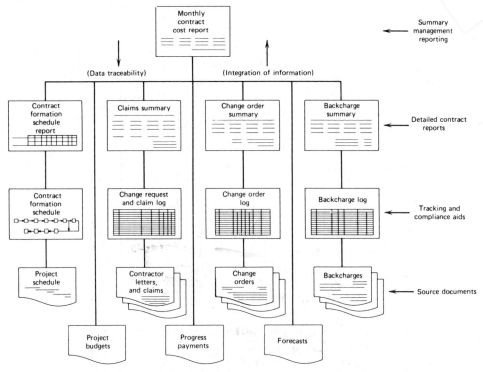

FIGURE 22. Contract reporting hierarchy.

leading to (4) a summary report for higher management. The documents, records, and reports should be so structured that data and information contained in each is integrated and that management can trace down the reporting system to understand the causes and effects of information it receives.

A monthly contract cost report that provides key information regarding contracting performance is shown as Exhibit 24. It incorporates and summarizes significant information shown on detailed contract reports, illustrated by Exhibit 25 (contract formation schedule report), Exhibit 26 (claims summary), Exhibit 27 (change order summary), and Exhibit 28 (backcharge summary). Each of these detailed reports is in turn fed by lower-level reports or source documents used for contract transactions. These should be continually maintained and filed with active contract records. Several of these source documents have been described and illustrated in earlier chapters. Depending on the structure of the project or company organization, detailed contract reports (change orders, backcharges, etc.) may also be submitted to management as part of a periodic contract or project report package.

EXHIBIT 24. Monthly Contract Cost Report

MONTHLY CONTRACT COST REPORT

Number of Contracts Awarded	_____
Change Orders Issued	_____
Backcharges Issued	_____
Unresolved Claims	_____

Total Original Contract Price	$ _____
Authorized Price Increases	$ _____
Accepted Backcharge Amounts	$ (____)
Expected Claim Amounts	$ _____
Total Contract Forecasted Cost	$ _____
Current Budget Contract Costs	$ _____
Cost Variance Expected at Project Completion	$ _____

Details of Contract Status:

Contract No.	Description	Budgeted Cost	Original Committed Amount	Change Order Amount	Back-charge Amount	Present Committed Amount	Paid to Date	Estimated Cost at Completion

EXHIBIT 25. Contract Formation Schedule Report

CONTRACT FORMATION SCHEDULE REPORT

Contract No.	Description		Specs and Drawings Received	RFP Issued	Prebid Meeting	Bids Received	Award Recom- mended	Contract Award	Commence Work
		Scheduled Date							
		Actual Date							
		Scheduled Date							
		Actual Date							
		Scheduled Date							
		Actual Date							
		Scheduled Date							
		Actual Date							
		Scheduled Date							
		Actual Date							
		Scheduled Date							
		Actual Date							
		Scheduled Date							
		Actual Date							

EXHIBIT 26.   Claims Summary

CLAIMS SUMMARY

Contract Number	Contract Description	Description of Claim	Claim Amount	Unresolved	Resolved	C.O. Amount	Remarks

Project _____
Period Ending _____

	This Period	To Date
Number of Claims Received	_____	_____
Total Amount Claimed	$ _____	$ _____
Number of Claims Resolved	_____	_____
Cost of Claims Resolved	$ _____	$ _____
Expected Cost of Unresolved Claims	$ _____	$ _____

EXHIBIT 27. Change Order Summary

CHANGE ORDER SUMMARY

Contract Number	Description	Issued C.O.'s		Pricing Method Used			Proposed/ Potential C.O.'s		Original Contract Price	Amended Contract Price
		No.	Amount	L.S.	U.P.	C +	No.	Amount		

Project _____
Period Ending _____

	This Period	To Date
Number of Change Orders Issued	_____	_____
Cost of Issued Change Orders	$ _____	$ _____
Number of Unresolved Change Orders	_____	_____
Expected Value of Unresolved Change Orders	$ _____	$ _____

243

EXHIBIT 28. Backcharge Summary

BACKCHARGE SUMMARY

Contract Number	Description	Description of Backcharge	Backcharge			Unre-solved	Amount	C.O. No.	Remarks
			Cost	Amount	Accepted				

Project _____
Period Ending _____

	This Period	To Date
Number of Backcharges Presented		
Cost of Backcharges Presented	$	$
Number of Backcharges Accepted		
Cost Recovered through Backcharges	$	$
Remaining Recoverable Cost	$	$
Remaining Unrecoverable Cost	$	$

KEY PERFORMANCE INDICATORS

Regardless of the reporting system characteristics, certain "key indicators" help management understand contracting status and performance. Depending on the amount of contracting used and the elements of risk involved, most of the following key indicators should be monitored.

1. Contract formation
   a. Number of qualified bidders
   b. Number of addenda issued
   c. Number of responsive bidders, compared with number invited
   d. Planned versus actual dates for formation activities
   e. Frequency of use of letters of intent
   f. Average effective period of letters of intent
   g. Percentage of contract amounts priced as lump sum, unit price, and cost plus
   h. Deviations from planned pricing methods

2. Contract administration
   a. Planned payments versus actual payments
   b Number of change orders
   c. Percentage of change order cost to original contract prices
   d. Backcharges (number and amount)
   e. Claims (number and amount)
   f. Percentage of change orders priced as lump sum, unit price, and cost plus
   g. Average elapsed time identification of need for change, change approval, issuance of change order, and payment
   h. Owner time charges to contract administration tasks

REPORTING CHALLENGES

Despite the best intentions, contract reporting often falls short of its objectives. Certain problems persistently plague the design and implementation of effective contract reporting and should be recognized and eliminated. A partial list of the most common challenges to contract reporting follows:

1. Timeliness of information
2. Consistent cutoffs in collection of cost, schedule, and performance data
3. Integration of contract data elements, and integration of contract data with other project information
4. Data traceability from summary reports, through detailed reports, to source documents

5. Reporting of potential problems in addition to approved transactions, such as disclosure of pending claims and changes rather than only those that have been resolved or approved

6. Exception reporting: that is, information demanding analysis or further management attention, and separating this information from the sea of data available

7. Reconciliation of estimated costs, such as when a cost-plus change order is issued, with actual costs when the changed work is completed and contractor charges are known.

8. Management attention: that is, getting owner and contractor management to evaluate and act on information presented in contract reports.

CHAPTER 19

CONTRACT AUDITING

To determine the efficiency of contracting processes and ensure that contract charges are correct, most owners and some contractors use some sorts of contract audits. Auditing differs from most of the other contract-related functions already described in that it is not directly required in order to conduct contracting business. Rather than being a specific set of activities that must be performed during the planning, formation, or administration of contracts, auditing represents an independent, arm's-length review of contracting processes as a whole or of specific transactions deserving particular management attention.

The purpose of each type of audit described in this chapter is to give management a view. In the case of cost audits, this view may be close and detailed—focused on specific charges to a specific contract. Operational audits, though, tend to give management a broader picture, a larger view of the contracting processes being undertaken. In either case, audits do not take the place of standard contract formation and administration activities and controls. They are performed in addition to the normal review and approval of invoices, time sheets, quantity-installed reports, acceptance testing, and all the other evidences of contractual compliance. These are all elements of the *execution* of contracting activities. The purpose of auditing is to verify that these are being executed as effectively as possible.

TYPES OF AUDITS

There are three broad areas of auditing related to construction contracts. Each is conducted differently with a different set of objectives.

1. *Financial Audits*. Financial audits are designed to ensure that financial information regarding the status of the owner or contractor represents actual conditions. Most large companies are audited yearly in order to provide the figures for financial statements regarding their fiscal position (assets, liabilities, debt, income, etc.). Usually performed by an independent auditing firm of certified public accountants, this type of audit leads to their attestation that the company's financial statements present fairly the financial position of the company and the results of its operations for the period. For most owners, this is a very limited review as far as contracting is concerned. In most cases, contracting charges and operations are beyond the scope of financial audits. Occasionally, because of the size or scope of an owner's or contractor's contracting effort, and therefore its impact upon financial position, financial audits do examine contracts to some extent. On the whole, however, these types of audits should not be relied upon to ensure that contracting is being performed correctly, or even to identify contracting deficiencies.

2. *Cost Audits*. Cost audits, on the other hand, have a direct application to many contracting elements. Their purpose is to test and verify charges made under a contractual agreement. As such, they involve detailed examinations of contractual billings. Their scope usually involves matching contractor charges to (1) the contractual terms and prices, (2) verification that the work was performed, and (3) evidence of costs, labor hours, or other resource expenditures to achieve that performance. By its nature, cost auditing is performed after the fact; that is, amounts that have already been paid to a contractor are verified for accuracy, applicability, and reasonableness. Sometimes it is performed as prerequisite to payment, but this is generally not the case.

3. *Operational Audits*. Operational auditing differs from the cost variety in that it examines *processes* that are being conducted or contemplated in the future. Operational audits look forward and seek to identify ways to improve contracting activities and controls for the benefit of the owner or contractor. Its goals are increased effectiveness, economy, and efficiency. Although costs are sometimes reviewed as indicators of contracting economy or effectiveness, operational audits focus their attention on business control systems and processes.

COST AUDITING OF CONTRACTS

Cost audits are performed by each contractual party when it feels that a separate check should be made of contractual charges, or when required under contractual terms and conditions. Their purpose is to verify charges, billings, payments, and/or receipts. They differ from the standard review that is made to verify the accuracy of charges when they are presented. This is a function of those responsible for making payments, usually construction accountants or the owner's accounts payable department. Cost audits provide an independent or second review of these amounts.

As such, cost audits should not be intended as a duplicative effort to that which is common business practice during the course of contract administration. Cost audits should represent more than just the checking of numbers that have already been checked or the review of invoices and statements that have already been reviewed. Instead, they should be targeted to specific high-risk areas. They should test charges or payments on a selective basis, and that selection should be guided by the degree of risk involved.

The conduct of all audits should be structured to a fairly well defined scope and work program. Cost audits are no different. Work programs that describe the step-by-step process of auditing, or the audit procedure, should be designed to identify errors and overcharges that are more apt to occur by virtue of the contractual pricing structure. Given the massive amount of data and records associated with payments under large contracts, it goes without saying that auditors do not have the time or personnel to review everything. The selection of charges to be tested is as important as the testing itself. For this reason, audit work programs should vary in scope and conduct, depending on the type of contract to which they will apply.

One would think that, particularly on large and sophisticated projects, this auditing would be planned and performed with great care. This is often not the case. Many contract audits, whether performed by an internal auditing group or an independent auditing firm, are merely superficial exercises. Typically, they center around a repetitive and mundane comparison of contractor invoices with associated cost substantiation and the contract itself. Much time and expense is spent chasing minor clerical or accounting errors through this process. This "shot in the dark" approach could be greatly enhanced if the audit detail and associated work programs were tailored to the risks inherent in each type of contract being audited.

It goes without saying that the risk of error, overcharging, and waste, from the owner's perspective, is greatest with cost-plus contracts and least with lump-sum contracts. Consequently, owner involvement and degree of audit detail should increase as the particular contract migrates toward a cost-plus agreement. But there are cost risks associated with any contract, regardless of its pricing structure. And each contract presents different elements of risk because of its particular pricing structure.

Let's look at the areas of common error and overcharge for each major type of contract and just a few of the corresponding audit techniques applicable to each.

Lump-Sum (Firm-Fixed-Price) Contracts

For lump-sum or firm-fixed-price contracts, the area of frequent error or overcharge is with change orders. Applicable audit techniques that help avoid problems with change orders are:

1. For selected contracts, verify that change orders represent changed or added work, that is, work not covered under the scope of the base contract or previously issued change orders for the contract under review or other project contracts.
2. Where specific change order pricing provisions exist in the original contract, verify that they were used. If not, suitable justification for the deviation should be documented.
3. Examine the time lag between identification of the need for extra work, such as a drawing revision, approval and issuance of the change order, and performance of extra work. Many change orders are issued after the fact, when the owner is in a poor bargaining position.
4. Match lists of issued, pending, and "in-process" change orders provided by the owner with independent lists furnished by project contractors. Surprising discrepancies often exist.
5. Examine the method by which change orders are priced. Was the most economical and controllable method chosen or are cost-plus terms being used more than necessary?

Unit-Price Contracts

For unit-priced contracts, the area of frequent error or overcharge is in the determination of quantities. Audit techniqes that help avoid problems with determining quantities are:

1. Match "quantity installed" reports with invoices.
2. Check that following completion of changed work *actual* quantity determinations were made for change orders that were based on estimated quantities when issued.
3. Investigate the need for independent verification of installed quantities through the use of as-built drawings, quantity surveys, and so on.
4. Verify that ancillary services and products normally included in unit prices, or specifically included per the contract, are not invoiced or paid for separately.

Cost-Plus Contracts

For cost-plus contracts, an area of frequent error or overcharge is in the determination and allowability of costs. Audit techniques that help avoid these problems with costs are:

1. Examine controls to ensure that cost-plus resources, including cost-plus change orders, are not being used to perform fixed- or unit-priced activities by the same or other contractors—for example, double payment for the same work.

2. Reconcile contractor time sheets with personnel brassing or other sign-in/sign-out records.
3. Verify allowability and reasonableness of general, administrative, and overhead charges.
4. Test allocation of contractor home office overhead to the project in accordance with contract terms.
5. Reconcile labor rates for construction craft personnel with actual local and national collective bargaining agreements in effect at the time the work was performed and/or contractor payroll records.
6. Check that construction equipment rental rates utilized were most beneficial to the owner; for example, monthly rates, rather than the more expensive weekly or hourly rates, used if equipment was rented for 30 days or more.

OPERATIONAL AUDITING OF THE CONTRACTING PROCESS

Major contracting efforts entail a great deal of cost, time, and personnel. From an owner's viewpoint, the construction of a new facility often equals or exceeds the cost of any other concurrent operation. Most advocates of operational auditing feel that all major company operations should be periodically reviewed to assure company management and owners that these operations are being performed with an acceptable level of control and to recommend improvements for the company's benefit. Operational audits, therefore, play an important role in the management of major contracts.

Operational auditing should determine (1) if systems are in place to ensure an adequate level of control over each operation, and (2) if those systems are being operated correctly. Operational audits, therefore, begin with an examination of the controls that presently exist over each of the contract activities described in this book. System deficiencies are noted and described. The second phase tests the operation of those systems that are in place. System and operating deficiencies are then noted, their negative impacts described, and the results reported to company or project management. Most operational audit reports include recommendations for both system and operating improvements that, if implemented by management, would reduce cost, liability, and inefficiency, as well as increase control over the processes themselves. The perspective of operational auditors, then, should be forward looking. They exist only to seek and recommend improvements, not to criticize operations or find fault with those responsible for deficiencies.

Many medium and large corporations have instituted formal internal audit capabilities. Specific auditing departments or groups have been formed to conduct both cost and operational audits, and often these groups enjoy reporting relationships directly with top management. They operate independently of production or line management, serving as the "eyes and ears" of company executives. This organizational position and independence should allow internal auditors the freedom to seek out system and operating deficiencies and unhampered ability to give top management an unbiased view of company activities and controls. It is not uncommon, however, for this position to be abused or underutilized. One of the greatest challenges of those charged with auditing responsibilities is to maintain their independence, their company-wide objective, and their working relationship with those responsible for elements under their review.

SUGGESTIONS FOR OPERATIONAL AUDITORS

It is impossible, except for the smallest jobs, to audit everything, but here is a short list of areas where audit attention might be well worth the effort.

Progress Payment Methodology.   Examine the detail of payment documentation and verification steps in comparison with the type of contract and risks involved. They should correspond. Check the time lag between commencement or completion of the actual work and the payment. Are there many cases of retroactive approval or payment for extra work? Are the forms, estimates, and so on, completed simply as an afterthought, where they fulfill documentation requirements but certainly do not evidence cost control? Are there cases where the detail and documentation are too severe, especially for low-cost or risk items—where some streamlining and flexibility should be allowed? (It makes no sense to spend a dollar controlling a penny.)

Retention.   Determine the actual cost of this practice and decide, in light of the risks involved, if it is reasonable. Are you withholding too much for too long, and subsequently paying higher embedded prices to account for this practice? If so, is the degree of control worth the cost and effort, or are more reasonable amounts and durations (or release requirements) in order? Also, are there other ways to assure satisfactory performance that might be simpler and more cost-effective?

Contract Documents.   Check for unenforceable, outdated, and expensive contract clauses. Determine what (if any) protection they bring, what they actually cost, how they might work against you, and the administrative requirements they entail. Look for "nice to have" but not necessary requirements that could be reduced or abandoned. Look for prerogatives that you insist upon, either out of habit or ignorance, that cause a lot of hassle yet buy you nothing in terms of risk avoidance.

As-Built Information (Drawings, Surveys, etc.).   First, determine if you need these and are not getting them. Then decide whether, for those you require, you actually need them at all. If so, how are they being verified and maintained? Are you paying for something you do not use or never receive? Are you receiving and maintaining something you will never need? On the other hand, are you neglecting to require these and maintain them as you should? Also check the methods by which their accuracy and authenticity are verified, and their quality. Are you paying for new drawings and getting old drawings with cryptic notes and scribbles?

Mobilization Fees and Other Overhead Charges. Evaluate the pros and cons of embedded overhead items versus paying for these separately. Check whether too many large, identifiable cost items are buried in these cost pools, ones that should be identified separately. Then do the opposite: check for minor cost elements that are billed and accounted for separately that could just as soon be included in overheads or mobilization charges. These might include supplies, consumables, inexpensive equipment, and the like. Make sure that more money is not spent processing and verifying these costs than the costs themselves.

Estimating Change Orders. Are these issued for a fixed amount, a ceiling, or "not to exceed' price and then never verified after actual costs are incurred and charged? For example, a common mistake is to approve, say, $10,000 for repair work, process the change order, and then never follow up to determine that only $5,000 worth of work was needed. Here, a "not to exceed" limit is transformed, by accident, into a fixed-price, lump-sum authorization. Contractors will not come to you after the work is done and ask to return money—but you can be sure they will appear if the work cost more than the "not to exceed" price.

Scheduling of Contract Formation. Are bid, evaluation, and award steps compressed into such a short amount of time that sound, market-competitive decisions are not made? Are shortcuts taken? On the other hand, are contracts awarded and payments committed too far into the future for no defensible reason? This brings a host of unnecessary problems: ties down your cash, ties you down to a contractor that might suddenly get busy or go broke before you need the work done, and subjects you to cancellation charges should the project be deferred or delayed before it begins. All are preventable risks.

Distribution of Contract Documents and Records. The more copies you distribute, the more you pay for revisions, and the more difficulty you have assuring that every recipient is working to the latest version. Do not give them to people who don't need them—and don't keep them from those who do. If there is any common error here, it is going overboard—sending too many copies to too many people.

Change Orders. Do not be satisfied by examining a sample set or testing each one on its own merits. Look for trends across change orders. Look for an excessive number in any one area (for example, in electrical work, or for a particular subcontractor) and find out why they occurred. Are they a symptom of poor design? Poor management? Sloppy scheduling?

Look at the pricing method for change orders and determine whether the cost-plus alternative is abused. Often it is, for it is the quickest way to get the paperwork settled and the work performed.

Building A Contract Risk Profile

Audits should be focused on risk—the higher the risk, the more detailed and frequent the audit attention. To help you determine risk and focus your effort, consider building a "contract risk profile" for each contract or subcontract. Use some or all of the criteria described here to rate each contract. You might even want to assign a ranking or risk factor to each criterion and calculate a weighted average to home in on those contracts with the highest risk (and greatest potential for audit payoff). Here are some of the criteria you might consider:

1. *Contract Pricing Structure.* This is pretty straightforward—the "softer" the price, the greater your scrutiny. Rank cost-plus work more risk prone than lump-sum, fixed-price agreements and changes.
2. *Change Order Profile.* Factor into your ranking the number, cumulative cost, use of hard-money versus cost-plus, and timing frequency (more frequent = greater risk). If there are large timing lags between identifying the need and issuing the change order, rank risk higher. Assign a higher risk in cases where actual costs of changed work are close to estimated costs. Are the estimates that accurate, or are the estimated costs that are approved abused? Include in your risk factor a component that accounts for disputes and discrepancies—the more of each, the higher the risk factor for the contract.
3. *Claims.* The more claim activity, the higher your risk factor and the more reason to focus audit attention on the contract. Consider a higher risk factor as the number, size, age, and assessed or committed liability increases for each contract you examine.
4. *Contract Life Span.* The longer the contract is active, the greater the chance for discrepancy and abuse.
5. *Upsets.* Rank contracts as higher risk if they have been canceled in midperformance, delayed, or awarded on a staggered basis.
6. *Pricing Mixes.* Be alert to the higher risk that comes with mixing more than one pricing technique for the same contractor. Look for cases where lump-sum, unit-price, and/or cost-plus are commingled. This is fertile ground for error and abuse.
7. *Postbid Negotiated Terms.* When significant terms and conditions, especially prices, are agreed upon after the contract is awarded, rate the contract as high risk.
8. *Design Revisions.* When these are numerous and frequent, risk goes through the roof.

9. *Personnel Movement.* If a key owner representative switches employment to the contractor, you have a potential risk involved.

10. *Dealing and Trading.* Look out for labor-hour swapping, where the owner provides some work that the contractor should have done, or vice versa, and side deals are made to reconcile all this later. Look for cross-charging, with or without the owner's permission. This often occurs when accounts are not yet open and the work must be done, when price thresholds are exceeded, and the work is not finished, or when record keeping is unattended. All of these cases indicate a lack of control and potential for abuse. In most cases, the problem is one of error rather than intentional malfeasance, but it can be a tremendous problem nonetheless.

11. *Growth in Owner Scope.* If the owner's scope of work, items furnished by it, or responsibility increases, rank the contract high risk in this area.

12. *Audit Aging.* The longer the period since the last audit of this project, contract, or contractor, the higher the risk—or the more reason to target it next time.

13. *Extreme Schedule Pressure.* When people and companies are facing deadlines, controls slip and cost can be disregarded. The higher the schedule sensitivity of a contract or portion thereof, the greater the potential for problems.

14. *Long-Term Relationships.* Rank contracts high risk if you have been doing business with the contractor or its representative for a long time. Familiarity breeds laxity of control and potential for abuse.

15. *Owner Management and Staffing.* If the owner is understaffed or staffed with inexperienced people, rank the risk level as high. If there is a great amount of turnover or organizational turmoil, do the same.

16. *Contract Management.* If experienced, competent people are handling the tasks described in this book, if records are well maintained, documents are up to date, planning is thorough—in other words, if contract management is being performed professionally, rank the contract(s) in question as low risk. This is probably the most telling indicator of all, the trust that you and your company can place in the people and systems responsible for safeguarding your interest. Auditing helps, and it can pay tremendous dividends, but if it follows poor contract management, it is not enough—it is *damage control*, not risk prevention.

THE ROLE OF AUDITORS

There are two distinct types of auditors that perform the three major audits described earlier: internal and external. Internal auditors work for the company they are auditing, whereas external auditors are temporarily hired from the outside. External auditors, such as independent public accounts, are generally responsible for financial audits, although they may be engaged to audit costs or operations as well.

The theory of auditing, its value and use, is often well received by most responsible owners and contractors. Problems encountered with audits do not generally involve the need for an audit or the authority to have one conducted. They do, however, center around the role, conduct, and suitability of the auditors themselves. Financial auditing is usually so far removed from the daily activities, interest, or expertise of most people who deal with construction projects that it interferes little with their activities. Cost and operational audits, on the other hand, often bring the auditor into close proximity with those responsible for performing the functions that are the subject of the audit.

The most pervasive criticism of auditors, be they internal or external, is that they do not know the business of construction or understand the complexities of construction contracting. This is generally a well-deserved pronouncement and one that auditors and their management should heed. Much has been said about the need for credibility and understanding of construction processes on the part of the contract manager. Auditors who will be involved with construction or contract activities share that need. The selection, training, and assignment of auditors is beyond the scope of this chapter. Let it suffice to point out that, particularly for operational auditors, actual functional experience together with formal training and exposure to the industry and its practices are essential. Provided that auditors have the proper training and experience to perform their duties, their success depends upon the willingness of company management to consider and act upon their recommendations.

One final recommendation is in order. Often owners station an auditor permanently on the construction site, expecting him or her to provide continuous monitoring of operations or costs. This is not recommended for several reasons. The primary one concerns the auditor's independence, or others' perception of that independence. Should he or she become deeply associated or involved with daily operations, or even become a part of ongoing execution tasks, this independence is lost. Another problem is that given the demands of contract administration or construction accounting responsibilities, the "independent" auditor is sometimes used to perform these activities. When this happens, the audit presence and effectiveness are diminished.

CONTRACT AUDIT CLAUSES

Most well-written construction contracts give the owner the right to inspect and audit the books of the contractor upon specified notice and subject to certain minor conditions. In addition, these contracts require the contractor to maintain accurate and complete cost records, particularly with cost-reimbursable agreements. It may be suggested that nonreimbursable contracts (lump sum or unit price) need not include such a clause. This could be true, provided that no instances requiring an audit or review of contractor costs arise. But there are many reasons why, even for lump-sum or unit-price contracts, auditing may be necessary.

Typical claim situations are an example. Even though the contractual setting may be fixed price, the very nature of claims is that they demand cost reimbursement for unforeseen events. Change orders performed under cost-reimbursable conditions are other examples of the need for adequate audit rights in all contracts, regardless of their original pricing methods. Wise owners make sure that such clauses are included in their general conditions and are careful that, when inserted, they in fact allow the types of review or audit anticipated.

Even the most precisely worded audit clause may be subject to vastly differing interpretations when its use comes into play. Most clauses specify the right to have records audited but are not as specific as to who may perform this audit. If "company auditors" are specified, does this mean the owner's external public accounting firm, or the internal auditors employed by the company? The "records" that are subject to review are also difficult to identify unless specifically described in the audit clause. Questions sometimes arise as to which records are meant by "records," whether these are only contract- or project-specific contractor records, or whether they are *all* records maintained by the contractor, even those pertaining to other contracts for other owners. Again, precise definition in the RFP and contract documents ensures that the owners' specific audit objectives are achieved.

Another issue that often surfaces when audit clauses are called upon concerns the right of the owner to audit the records of subcontractors or suppliers to the contractor. This right, if desired, should also be specified in what are called "flow-down" clauses—those that require the contractor to enforce similar clauses or obligations with its subtier agreements for the benefit of the owner.

AUDIT AWARENESS

The frequency of internal and external audits of all types is increasing among construction projects. This is due in part to the growing number of interested parties involved in the effort. It is becoming more common to have more than one owner financing a project and more than one contractor performing work. Joint-venture agreements also increase the need for separate auditing to protect the interest of each partner. Closer outside scrutiny by the public, corporate stockholders, special-interest groups, and the like have increased the demands for accountability of those responsible for construction contracting. All these factors have combined with economic and commercial pressures to expand both the role and frequency of construction audits, and this growth shows no sign of subsiding in the near future.

Because of the increasing scope and frequency of audit efforts, those responsible for contract management functions should evaluate each of their activities from a preaudit perspective. That is, they should plan and conduct their affairs with the expectation that they will be audited to one extent or another. This should cause a greater emphasis on thorough documentation of alternatives, decisions, and business transactions that occur during contract planning, formation, ,and administration. Forms, documents, records, and reports should be designed, used, and maintained to provide a clear audit trail. This "audit awareness" should not interfere with contracting activities by causing undue delay in decision making or by intimidating contracting personnel. On the contrary, it should achieve two interrelated objectives: (1) audit time and expenses will be reduced through clear audit trails, and the difficulties arising from the perception of lack of control will be eliminated; and (2) the knowledge that activities and transactions may be audited in the future serves to increase their efficiency and effectiveness for the immediate benefit of all concerned.

PART 5

SPECIAL RECOMMENDATIONS

This part contains two topics that apply to all phases of contract management, and depending on your project responsibilities and setting, may or may not be directly applicable to your contract responsibilities. In any case, exposure to the special principles in this part will help round out your understanding of contract management and sharpen your skills should you confront these particular challenges in the future.

Since contracts are legal documents and most contracting efforts involve legal rights and responsibilities, not to mention disputes and lawsuits, almost everyone concerned with good contract management should know how to choose and use competent legal counsel. That is the subject of Chapter 20.

If you do any work outside your own country, have suppliers, contractors, joint-venture partners, or others from other countries on your project, or simply want to know how contract management varies in the international arena, you should be interested in Chapter 21.

CHAPTER 20

THE ROLE OF A LAWYER IN CONTRACT MANAGEMENT

THE RELATIONSHIP BETWEEN CONTRACT MANAGEMENT AND SOUND LEGAL ADVICE

If you are involved with construction projects of any consequence—in any capacity—you will need the counsel and skills of a lawyer. Although it is impractical to suggest that every contract manager or every professional involved in the process have a license to practice law or even legal training, all of us need to know three things: (1) when to use the services of legal counsel, (2) how to choose a lawyer, and (3) how to use one once you have made your choice.

The first question can be answered very directly. Use a lawyer from the very beginning, because that is when the mistakes begin. Continuity—the involvement of one person or firm from the early stages all the way through to project completion—is always a plus. There are, however, mileposts in the contract management sequence that demand particular legal review and advice. This chapter describes those risk points, offers advice on choosing a lawyer, and gives suggestions for making the most of his or her expertise. For every side in every contract, then, it is not a question of *if* you should enlist legal expertise, it is *when* to use it, *who* to use, and *how* to get the most from the expertise you are buying.

How to Choose a Lawyer

No engineer would specify an inferior product or select alternatives without consideration for their usefulness and quality—and their price. The same holds true for choosing a lawyer. Here are the major factors that you should consider when deciding which firm and person should be selected.

1. *Practice Type.* Hire a lawyer just as you would an architect, construction manager, or any other employee—find one who knows the business of construction, the problems and risk it entails, and the solutions for them. In other words, *find one who understands construction disputes.* Lawyers specialize in all sorts of fields, and one perfectly qualified in one may be a handicap in another, particularly if the other is your field—construction.

Do not look to one lawyer for everything. The law has become too sophisticated for one person to handle it all. If you will have corporate as well as litigation needs, consider a large practice with specialists in multiple practice areas. If you are only concerned with a single project or a particular dispute, choose the specialist in construction regardless of his or her firm.

2. *Personalities.* You may end up spending many hours with your lawyer, under high-tension situations, and under less than favorable circumstances. Interview several suitable prospects and choose one you like. Make sure he or she understands that *you* are managing the contract, and that he or she respects your profession or trade and is familiar with the stress and practices you face.

3. *Price.* The legal market is increasingly competitive. If you have a large amount of legal work to be performed, the chances are that you can negotiate a fee agreement more favorable than is standard for the lawyer of your choice. Always remember, however, that the purchase of legal services is no different than others: You get what you pay for. Don't try to save a few dollars on attorney fees when to do so may jeopardize your project, your financial position, or your business. Consider contingency fee arrangements that are outcome dependent. This is especially important if your legal bills are predicted to cause cash flow problems. Shop around—this is done for other professionals, why not lawyers?

4. *Utilization Strategy.* Once a law firm has been chosen, direct your requests to the least expensive lawyer who is competent to handle the work. For most services, lawyers charge on an hourly basis, and the more experienced, more senior professionals naturally charge a much higher rate—even for the same work. Do not make the mistake of specifying a sledge hammer when a tack hammer will do the job—look for the least expensive lawyer with the competence you need. If you are not acquainted with the law firm, this might be easier said than done. You will probably have to rely on the lawyer responsible for your matters to assign the tasks to other lawyers as he or she sees fit. But once you know the lawyers in the firm, try to identify competent ones who charge lower billing rates.

There are some legal tasks that take the same amount of time to perform regardless of experience level. If these are performed at half the billing rate of the most experienced lawyer in the firm, you will cut your bill in half. Often, competent younger (less expensive) lawyers can spot all the issues and get the advice of more experienced lawyers by asking a few questions, and even for free—in casual conversations or shoptalk. Make sure that the younger lawyer has enough sense to find the correct answer—that he or she is not afraid to ask questions of superiors. Make sure that your lawyer understands his or her limitations.

5. *Expenses.* Expenses and billing habits of law firms are often nego-tiable, and they can greatly reduce or increase your cost, depending on how they are handled. Most firms mark up and bill expenses directly related to client matters for such items as phone calls and photocopying. To cut these costs, consider handling some tasks outside the law firm—by doing them yourself or with less expensive help. Examples might be large photocopying tasks (lawsuits often involve extreme amounts), interviews, document searches, and the like.

Make sure that you understand expense terms of all types, to avoid being hit with a surprisingly large bill. Know in advance which will be passed on at cost and which will be marked up—and how much. Few things interfere more with client–attorney relations than disputes or misunderstandings of this type. As with all other forms of contracting, use your negotiating strengths as a buyer to obtain the most favorable expense terms.

You may be able to convince a firm to alter its billing rates and habits on your work. For example, you may reach an agreement that the lawyers will not bill for travel time. In a far-flung case, this seemingly minor concession may save you a fortune.

Finally, always consider how you are being charged and use that knowl-edge to your advantage. If you are paying by the hour, do not spend time chitchatting with your lawyer or discussing unrelated topics during the busi-ness day. When determining what time to bill, the lawyer may not be able to separate casual conversation from substantive, billable work. Always know when the legal meter is running. The time is expensive, no matter whom you hire. Use it wisely.

THE LAWYER'S ROLE DURING CONTRACT FORMATION

Competent legal advice *during contract formation* is the best bargain in the construction industry. Seldom are fewer dollars spent to gain more value. Remember this: The purpose of the contract(s) you are forming is not to construct a building or a highway, or to construct anything. *Contracts don't build,* and building alone does not require contracts. The purpose of contracts is to *establish and define the legal rights and obligations* of each party involved. Contracts allocate risks, risks ultimately founded in the legal rights of the parties. Without a valid contract, the owner has no protection, except as established by statutory law and equity. Statutory law is generally under-stood thoroughly only by lawyers, and equity is established in a court of law. Reliance on either, or both, is a mistake. Both are inadequate from the standpoint of all concerned.

Legal Guidelines for Formation Tasks

Like a stage play or a novel, a discussion of legalities needs a setting. This section assumes, except where specifically noted, a setting where the owner has some recognizable title to the property on which the project is to be built. For example, ownership in *fee simple,* ownership *subject to a mortgage,* or ownership with legal title transferred under a *security deed.*

These settings are the most common on complex projects and are distinguished from those where the property is owned by the contractor during construction and then transferred to the owner under a deed, such as when a home builder finishes off a "spec" house to the specifications of the buyer. This may seem like a trivial distinction, but without it, the impact of arcane, unpredictable, and even bizarre real estate law may be pervasive. The "owner" in this context is a somewhat sharper distinction than the term as used in preceding chapters. Here "owner" means the owner of the real estate on which the project is being built, and for whose direct or indirect benefit the construction work is being performed.

1. *Suggestions Concerning Law When Forming Contracts.* Most contracts are written or reviewed by lawyers but are implemented by nonlawyers. However, most disputes rely on what lawyers, and courts, read in those contracts. The challenge of contract formation is to write contracts that are understandable and enforceable, and that help solve problems rather than create them. Have an attorney review your standard contract forms or ones that you have altered for each application. If you are a contractor or subcontractor given forms created by others, get them reviewed by competent legal counsel. Discuss the risk allocated to you under the proposed contract. Besides that general advice, here are some specific tips:

a. *The law is always changing.* What is legal today may be illegal tomorrow. A contract clause that was once enforceable may now be worthless and void. A procedure once insufficient to protect the owner may now be sufficient, and vice versa. If you do not know the impact of the law, your contract does not provide much guidance from the outset, and it could be your nemesis during the administration period. Ignorance may be bliss for some, but according to the courts, ignorance of the law is never an excuse. Do not be satisfied simply because the resident engineer or the contract manager knows the contract. They are not always the only ones who implement contract provisions or jeopardize your rights under them. Make sure that the rest of the staff—the people who are working directly on the project—know the contract conditions.

b. *Always read your contract* before *deciding to use it.* It does not matter if you or the other party originated the contract—when you sign it, it is yours. The law makes it the duty of a party to read every contract it signs. This sounds simple, but many problems arise when people assume they know the contract terms when they actually don't.

c. *When reading, ask yourself, "Do I understand this?"* The law holds ambiguous and self-contradictory contract language against the party that proposes or submits the language. Never propose a contract with ambiguous language.

d. *When reading, ask yourself, "Will contractors understand this?"* Remember, the documents are the script. If the script is deficient, the work stands a good chance of being defective. It does not suffice if only the owner understands the contract—the contractors and their subs are the ones who will be working under it. Avoid in-house slang, national idioms, and wording that can be interpreted more than one way. Keep it simple and direct. You are after a completed project, not to impress anyone with your mastery of language. You win when you are understood—not when the other party is confused or impressed.

e. *The written word can be trash or treasure.* Know the difference, and if you don't, hire a lawyer who does. Most standard or reference contracts were drafted or modified by lawyers at some point. There was a time when lawyers wrote only in "legalese," but a recent trend is toward easily read contracts that revert to "legalese" only where necessary. Sometimes legalese is meaningless, sometimes it means something that can be better said in everyday language, and sometimes it can give one party tremendous leverage over another. The challenge is to determine what each instance of legalese means and whether you want it in your contract. These guidelines might help you make that determination:

(i) If a standard or "form" contract seems to be all legalese, it is probably old, outdated, may never be understood by the parties, and should not be used.

(ii) If a form contract is generally understandable, but seems formal, has occasional insertions of legal language (like "waiver" and "estoppel"), and refers to a lot of paragraphs and subparagraphs, it probably was recently drafted by lawyers. If you do not like the legalese, don't change it without the assistance of a lawyer. Amateur legalese can be much worse than pure legalese!

(iii) If a contract is generally understandable, seems to have occasional legalese, but contains inappropriate provisions or references, it is probably a form that has been changed over the years for application-specific purposes. Unless you are repeating the identical project, with the identical parties, under the identical risk circumstances, this contract could spell disaster in a dispute or lawsuit. Don't use it! Start with a new, unrevised form contract, modify it for *your* use, or see a lawyer.

(iv) If the contract is very clear and easily understandable by anybody, it may get the job done—but it probably does not provide the protection and controls you would like to have on a large and complex project. Remember: a contract allocates risks. If you do not allocate the risks to another party in a contract, you are by default taking the risks upon yourself.

An extreme example of this type of contract is the form "proposal" submitted by a home improvement contractor, which typically states: "I will provide _____ for $_____." A signature blank is provided for owner acceptance. You cannot get much clearer than that, and you cannot complain about legalese. But this simplicity has a terrible hidden price—it leaves the owner totally at risk if it pays in advance. It omits controls over payment terms, completion schedules, and quality. The complexity of the work may be an indicator of the contract complexity required, but it is not a true one. It can be very bad business practice to assume that simple work demands simple contract provisions. If the contract is too simple for the complexity of the work *and the risk involved,* do not use it.

f. *Be careful with incorporation by reference.* Most commercial form contracts use a legal format known as *incorporation by reference* as a method of segmenting construction contracts into shorter, more coherent pieces. For example, the AIA General Conditions, 1987 Ed., has a special paragraph that lists the "contract documents" (See Chapter 5). These usually consist of the specifications, special conditions, and supplemental conditions. Since the totality of the documents comprise the contract, you are headed for trouble if the terms in any two documents conflict (causing ambiguity). Finding these conflicts is difficult enough if you originate all the documents yourself. Finding them when you incorporate entire documents by reference can be unbelievably trying.

g. *Protect yourself.* As optimistic as either or both parties may be at the time of award, every contract setting contains unique and virulent risks. Protect yourself early, in writing, by assuming that these risks will materialize. Use all the self-preserving clauses you can, and do your best not to relinquish them during negotiation unless you have to or want to for some other gain. This leads us to the question of "fairness."

h. *Don't worry too much about fairness.* A contract manager is paid to represent his or her employer or client, not to referee a construction dispute. Your job is to get the project built on time, within budget, and without compromising quality. If you take the initiative in contractual dealings and keep it, you can elect whether to fulfill your sense of fairness when an occasion arises.

Construction contracts are not like consumer transactions. Generally speaking, any contract that is not illegal can be enforced—no matter how harsh. Terms need not be "fair," just legal.

For example, the owner can require that the general contractor use subcontracts with "pay when paid" clauses. In some states, a well-drafted pay-when-paid clause in a subcontract allows a general contractor to use nonpayment by the owner as a defense to a claim for payment by a subcontractor. On their face, these clauses protect the general contractor. However, in some states, Georgia for example (currently—remember: the law always changes), when these clauses are properly used, an owner can prevent a subcontractor from perfecting its mechanic's lien rights against the owner's property by not paying the general contractor for the subcontractor's work. This is because Georgia law requires the subcontractor/lien claimant to get a judgment against the primary debtor (the general contractor in this case) as a prerequisite to the establishment of the lien. In other words, when the general contractor has the nonpayment defense, the subcontractor cannot obtain a judgment against the general contractor. Unless the general, in turn, successfully sues the owner for the payment, the subcontractor is left out in the cold.

Most reasonable people, and all subcontractors, would probably call this "unfair"—they would say that the general owes the subcontractor whether or not the general is paid by the owner. But whether something is fair or unfair is irrelevant in this case. The law, not a sense of fairness, controls the rights of the parties. If you, as contract manager or other representative of either party, decide to accept what you feel is fair, or to relinquish your rights because exercising them is unfair, you may be doing your employer or your client a disservice.

2. *Investigating the Financial Position of a Bidder or Contractor.* Lawyers can help here, too. The financial condition of a contractor is important because, under many contracts, the project is in effect funded by the contractor until final payment is made. This seems contradictory—isn't the owner supposed to fund the work? Well, not necessarily so. The contractor is actually funding the work when its profit margin is smaller than the retention percentage withheld, when payment terms are heavily backloaded, or when there has been a substantial time lag between the submittal of an application for payment and the date when payment is actually made.

From the owner's perspective, the contractor is the *financial barrier* (or buffer) between the owner and subcontractors. Even with lien claims a solvent contractor will be required to pay its creditors, directly or by indemnity, *before* the owner pays the contractor (to prevent the owner from paying twice). From the owner's perspective, with a proper contract, a *solvent* contractor is a source of protection against claims of lower-tier subcontractors and suppliers.

For these reasons alone, a contractor should be asked to disclose its major assets and liabilities. Prudent owners go further, depending on the circumstances, and require each contractor to identify pending lawsuits against it, and to reveal whether and when it (or another company owned or controlled by the same people) has ever filed for bankruptcy. Lawyers can help track all this down and help you assess financial risks and legal concerns that surface.

Recent events point to increased risk with sureties, and prudent owners not only insist on bonds, but verify that the bonds they get are worth something. If the surety is not solvent, it is no good to anyone. Do not rely on regulatory agencies to assure the soundness of sureties, either. Sometimes these watchdog agencies don't watch very carefully.

3. *Getting Information from Public Records.* Public records contain vast amounts of information. But lawyers are often the only people who know where to look, how to look, and how to break the "hidden codes" contained in the documents. This section discusses some common resources lawyers use and what they can reveal at little expense—using systems they work with everyday.

a. *Computer Databases.* An incredible number of databases exist. Many of them are available to everyone but cannot be cost-justified because of limited use. Law firms, which thrive on information, often have tremendous access to the world's database sources. They pay the fixed subscription fee and charge you only for services you require. The cost can be minimal.

b. *Corporate Filings and Registrations.* Every state in the United States requires all corporations to make filings with its respective secretary of state. A quick check will tell you if the corporation exists at all. If it does, the filing could tell if the contractor has "good standing," if it has ever been dissolved, and provide the addresses and names of the officers. Any peculiarities should be investigated because they may indicate chronic financial or legal difficulties.

Many states and countries require foreign (or out-of-state) contractors to post bonds to secure the payment of taxes and for other security reasons. If this is required and the contractor has not complied, find out why.

c. *What Court Records Can Reveal.* Here are three sources of information commonly found in federal, state, or local records.

(i) *Grantor/Grantee Tables.* The last thing you want is a contractor that does not pay its creditors. Grantor/grantee tables in the property records where the contractor's previous job was performed will indicate how many liens were filed on the job, and whether they were ever paid off or settled. It is not unusual for some liens to be filed on every project, but indications of excessive claims should be viewed seriously.

The real property assets of the contractor can be found here, too. Check for discrepancies with the contractor's stated assets. For example, does the contractor lease the building it claims to own? If so, investigate further.

(ii) *The Civil Docket.* You may find recent lawsuits in which the contractor has been involved here. Have your lawyer check in the court of proper venue (usually, the contractor's home base) and where recent projects were performed. The pleadings filed in these lawsuits can be very revealing. Pay special attention if large contract claims are brought or if fraud or other illegal practices have been alleged.

(iii) *Uniform Commercial Code Filings.* Every time a contractor gets a secured loan, the responsible creditor will file a statement (on county records in the United States) indicating that it holds a security interest in the contractor's collateral. Although these statements do not necessarily indicate that a loan was made or that debt is currently outstanding, they can lead you to the creditor named on the statement, so you can find out directly. Be wary if all the listed assets of the contractor are pledged as security for loans.

d. *Publications.* There are databases that contain years and years of articles printed in major newspapers, magazines, and other periodicals. You or your lawyer can search for news about the contractor by using its name in a search request. You can also search for articles that mention the principals of the contractor. Sometimes these lead to extremely pertinent information.

e. *Federal Government Records.* Two types of records in this category may prove informative: filings with the Securities and Exchange Commission, and federal tax filings.

(i) *SEC Filings.* All publicly held and many privately held companies in the United States are required to make certain disclosures to the Securities and Exchange Commission. These disclosures are wonderful resources of information concerning financial stability, including annual reports. These reports, though often hedged, must meet certain disclosure requirements. SEC filings may be obtained on computer databases.

(ii) *Tax Filings.* Many experienced owners know that the first creditor a contractor stops paying when it is in trouble is the Internal Revenue Service, especially regarding payroll taxes. This is probably because the IRS is slower to act than other creditors and will not stop jobs as quickly. Working with a contractor who has these difficulties is asking for trouble.

Many owners, or contract managers for that matter, never consider undertaking the types of investigations suggested here. Maybe that's why a lawyer is so helpful. He or she has been on the receiving end of the problems they are designed to prevent, and so has a special sensitivity—not to mention skill and access to records—that will pay dividends if used wisely.

THE LAWYER'S ROLE DURING CONTRACT ADMINISTRATION

Contract administration is the responsibility of the owner and contractor, and unless problems arise, lawyers are not directly involved. But since disputes and changes are common, if not planned, and since these often involve legal rights and remedies, lawyers are called in for specific, as-needed assistance.

Ascertaining Rights and Remedies

In a perfect world the contract should provide a lawyer with the necessary information to determine a legal remedy to most situations. If the contractor defaults, for example, the lawyer will recommend that the owner send the default notice provided by the contract, then terminate the contractor or exercise other available remedies under the contract, such as a backcharge. In this type of situation, the lawyer is merely validating existing contractual conditions for the owner—not a very trying task. However, actual situations are usually not as straightforward as this example.

Many times the parties' actions have deviated too far from the contractually required performance, and some terms of the contract may no longer be enforceable. A lawyer may be able to salvage the situation by redirecting the parties to performance under the contract, and thereby reestablish the enforceability of the contractual terms.

Lawyers, especially frequent litigators, are trained to make their clients' actions and positions look proper and reasonable—in case the dispute is eventually tried by a court or arbitration panel. If the *gloss* on the facts can be changed ever so slightly to the benefit of the lawyer's client, tangible advantages can be achieved. This process is sometimes referred to as *couching the facts*.

Take the example of a lawyer who can make the obvious bad-faith actions of his client look more reasonable and thereby avoid an award of punitive damages or attorney's fees. Seemingly small efforts, such as a tactfully drafted letter sent over the client's signature, can have tremendous evidentiary impact in a subsequent tribunal hearing the dispute.

Claims Assistance

Whether you are pressing a claim or defending against one, legal advice is critical. It begins with preparation and continues through the resolution process, and as with all aspects of contract management, should begin as early in the sequence as possible.

This is especially true for claims documentation and evidentiary materials. Keeping these current during the course of the project is far simpler and more effective than reconstructing them later. This advance preparation can even serve to reduce the number of claims that may ultimately be litigated, and gets you ready for battle when it comes.

Experience shows that the mere knowledge by contractors that the owner is well prepared to support its positions may dissuade even legitimate claims from ever being made. Sound preparation increases the contractor's fear of losing not only the claim but the substantial investment in expenses and attorney's fees litigation entails. In truth, many contract claims are either brought or dropped on pure cost/benefit analysis—not the basis of legal rights. In other words, claims proceed or are abandoned on the basis of their proposed cost and chances of success, not necessarily on who's right and who's wrong.

Legal Assumptions versus Fairness: An Important Distinction

Here we must return to the distinction between legal principles and perceived fairness. Remember, legal principles at hand may make assumptions that are not necessarily borne out in reality. Basic assumptions in a commercial construction context are that all parties are commercial actors (not individual consumers) that have equal bargaining power, possess the requisite business knowledge required by the trade, and have similar or equivalent access to the counsel of an attorney. Let's examine just two of these assumptions.

1. *Similar or Equivalent Access to the Counsel of an Attorney.* If, for example, the contractor operates without legal knowledge or competent advice until litigation starts, the owner can obtain technical legal advantages that may bar future claims by the unwary contractor. Although the assumption is that both parties have similar or equal access to the counsel of an attorney—one can take advantage of that counsel while the other does not. Let's look at an example.

Example. A drug distribution company was building a large warehouse. One of the multiple prime contractors was required to build a sophisticated fence within the building for the purpose of securing controlled substances. This fence was to span from floor to ceiling, attached at both ends. Special locks and conveyor belt access hatches were called for by the specifications. The contract documents provided that the fence would be completed approximately three weeks after commencement of work.

After reviewing the specifications, the contractor confirmed the firm-fixed price with the owner, who immediately sent the contract to the contractor for signature (execution). The contractor failed to return the signed contract and being in a hurry to get the work done, the owner allowed the work to start anyway. The contractor eventually completed the work, without ever signing the contract.

The contractor subsequently asserted a large extra work claim based on allegedly defective specifications and acceleration. The contractor's time schedules showed labor costs in excess of the contract amount, and the contractor asserted oral agreements with the owner's representative that all extra costs would be paid and that the contract was actually a time and materials (T&M) contract. The contractor sought twice the original contract amount.

The owner responded that the work was incomplete and required completion by a different contractor in order to meet the inspection date. The owner offered to pay the original contract amount, less the backcharge, but only if the contractor would sign the original contract document. The owner claimed that this was necessary for accounting purposes according to its established procedure. The contractor signed the contract and was paid the original amount.

When the contractor sued for the extra work, the owner raised a technical legal defense: the *parole evidence rule*. The contract provided above the signature line that "All negotiations and agreements made prior to the signing of this document are void."

The parole evidence rule is a substantive contract law that excludes any evidence of prior oral agreements by the parties to a later written contract contrary to the terms of the written contract. Clearly, the contractor's claim that the owner had agreed (orally) to pay the contractor on a time and materials basis was contrary to the terms of the executed firm-fixed-price written contract document. The court held that the contractor could not submit any evidence of the alleged oral agreement. This ruling destroyed the contractor's claim.

2. *Equal Financial Bargaining Power.* Owners often limit contractors' remedies by proposing the execution of carefully drafted documents such as change orders, lien waivers, releases, or even checks with language such as "final payment" on them. Contractors can be so anxious to receive payments that they will sign virtually anything to get them. This is understandable, for many contractors are not financially solvent enough to perform without current payment. So here again, a legal assumption, in this case "equal financial bargaining power," is incorrect. The gap between the legal assumption and commercial reality may be exploited to one party's advantage.

Handling Liens

For some reason, many owners consider real estate records to be sacred—and see a lien as a horrible stain on those records. This isn't necessarily true. Liens can be minor nuisances and easily handled, provided that you know how. Here are some tips that will help.

1. Never pay the lien of a subcontractor or supplier without consulting a lawyer. There may be times when this advice must be ignored, such as when you must open a business or sell property quickly and the lien is a minor defect holding up a major business transaction. But even in these cases, a quick legal check is easily obtained.

2. Whenever you do pay a lien, pay by joint check to the general contractor, subcontractor, and claimant involved. Otherwise, your payment may not count as credit toward the contract sum with the general contractor.

3. Most contracts require the general contractor to keep the property free of liens. If this is the case in your contract, force the contractor to bond off the lien. This frees the owner completely.

4. It is difficult for a claimant (contractor placing a lien) to *perfect* a lien (to make it legally acceptable and enforceable). Ask your lawyer to analyze the facts under a "lien defense" checklist to make sure that the claimant is in compliance with all technical requirements of the lien statutes. One mistake can be fatal to the lien. Make the claimant complete all of the steps under the statutes to perfect the lien—absent final trial in court. Aggressive attorneys will suggest that owners pay only when they know the claimant will recover at trial, unless the expense of the suit will outweigh the amount of the lien, and given a fair assessment of the lien claimant's probability of success. Even in this circumstance, the owner can negotiate for a discount.

5. If the lien causes a default under a loan the owner has with a lender, ask the lender to waive the default until the lien is verified as enforceable. Pursue remedies such as requiring the contractor to bond off the lien before paying the money, or if it makes business sense, bond off the lien yourself. Always seek indemnity from the general contractor where the remedy is available.

When the Owner Gets Sued

Unless the claim is minor, there's only one rule to follow in these instances: *Hire a lawyer*.

Litigation rules are governed by statutes known and understood only by lawyers. These rules are a complicated morass of practical structure and Byzantine dogma, all overshadowed by constitutional considerations learned in years of study at law school. Furthermore, the actual practice of the rules varies from their facial meaning. Except in the simplest cases, such as a suit on an account, sophisticated legal knowledge is required to prevent total and complete loss of the law suit.

When the Owner Decides to Sue

Even with a well-managed project, disputes and claims can resist reconciliation. If you think a suit is called for or imminent, take an aggressive role—sue first. Many lawyers believe that the posture of plaintiff is superior to being a counterclaim plaintiff. In either case, always view a lawsuit as an investment. Use a cost/benefit analysis to decide whether it is a wise one, and consider long- and short-term effects before proceeding. Litigation is not an embarrassment or a sign of management failure—it is simply a tool of business. Use it when it suits your purpose.

SUMMARY: GETTING THE MOST FROM A LAWYER

Regardless of when or how you use a lawyer, your actions and attitudes can become his or her greatest asset or biggest hindrance. For best results, help the lawyer all you can in obtaining and understanding the facts—all the facts, with nothing held back. Organize materials and documents, and compile data to make legal review fast and accurate—and less expensive. This will save you a load of money in legal fees and free the lawyer to do what he or she should: practice law, not manage the project or the contracts.

In this final respect, hiring a lawyer is no different than awarding a construction contract. Pick lawyers who are qualified, specify your needs in advance, agree to performance and payment terms, bring them on board at the right time, and let them do what they are hired to do.

As with other contractors, pay them for their work and give them the information, access, and freedom necessary to perform in your interest. Always remember that lawyers offer advice and representation—they do not manage your business or relieve you of your responsibility. Nor can they correct your management mistakes.

Contracts are legal agreements entailing rights and responsibilities. So it follows that most contractual problems are, by definition, legal problems. If you are serious about problem prevention and resolving problems that are unavoidable, you've got to know what lawyers can and cannot do for you. You've got to know when and how to use them. You've got to know the role of lawyers in contract management.

CHAPTER 21

CONTRACTING ACROSS BORDERS

Large construction firms that work in foreign lands are not the only ones that contract across borders. When an owner uses the services of a foreign vendor, contractor, or design firm, it is contracting across borders. The same is true for joint ventures where more than one nationality is involved, or when the owner's company is foreign-controlled, funded, or managed. It happens when people such as engineers or technical representatives come from other countries to work on your job site, you go to theirs, or goods, material, or equipment are shipped from one to the other.

In some countries and industries, this kind of complexity was once rare—but no longer. As the world's economies become more interdependent, as technology lets us transfer information and expertise worldwide in an instant, and as new markets, new companies, and in some cases, new countries open for business, cross-border contracting becomes more common. Indeed, for a project of any scale, it is hard to imagine a situation where all the required material, labor, knowledge, and money originates in one country and stays there. Contemporary projects of even modest scale are apt to involve contracting across borders, and this has two important implications. First, risk and complexity rise, sometimes quite unexpectedly. Second, and as a result, sound contract planning, formation, administration, and management are even more critical for everyone and every company involved. Contracting across borders puts a heightened emphasis every phase of the process, on every document, understanding, and practice.

It is not possible in one chapter to describe all the regulations, traditions, laws, and cultures that you might encounter in these situations. The intent here, however, is to point out general guidelines for consideration, reminders that cross-border contracting, be it buying or selling or both, is more challenging and carries much more risk than the domestic version. Murphy's Law translates into every foreign language and works in all cultures. It pays to be especially sensitive to "what can go wrong," because if you're not, it probably will. It also pays to know what you do not know, to question all your contract management assumptions and "understoods," because these can lead to tremendous problems. And it points out the need to use experienced professionals—attorneys, accountants, architects, engineers, and construction managers, to name a few—who are particularly sensitive to the perils and nuances of the work in question.

Another rule of thumb is that cross-border contracting generally takes more time, costs more money, and involves more steps than does work done in one country. Yes, bureaucratic barriers are coming down in some areas, as with the 1992 economic unification in the European Common Market, free-trade agreements between the United States, Canada, and Mexico, and the like, but they are still there to contend with. So are more subtle differences in culture, management styles, negotiation techniques, labor relations, and so on—those "understoods" that are not understood anymore spring up from the most unsuspecting sources and cause the most unexpected difficulty.

That is the general rule for contracting across borders: *Take nothing for granted.* Question every assumption. Do not accept implications—get your intentions expressed and your expectations fully understood. Do everything you can to learn about the particulars: laws, for instance, or work practices or technical standards, but never assume that you know them all. That is when they jump up and bite you.

In this chapter we address a whole host of issues that are subject to change when you, or they, contract across borders. Take these remarks as a sampling of the challenges but by no means an exhaustive one. Take them as general precautions, not insurance against mistakes or surprises.

What's So Different?

The shortest answer to this question is: *everything,* and most dangerously, those things that you think are the same. Contract law differs; professional registrations differ; taxes, currency, inspections, working days, religious customs, equipment specifications, union considerations, maintenance discipline, and languages all vary from place to place. Let's look at some of these from the point of view of "things that can go wrong" or issues to investigate and clarify before and during your contract experience.

Currency. In what currency will payments be made? Are currency exchange rates to be fixed in the documents or pegged to some index? When payments are made, is the current exchange rate used, or the one effective at the time of contract signing, when the work was accepted, when the payment is made, or some other time? All this needs to be specified in the documents and monitored closely, for there is a great amount of profit or loss to be had in creative currency exchange management. Some contractors have been known to make more on currency fluctuations than the work itself—and to lose more as well.

These conditions are difficult enough to establish and maintain when convertible, or so-called "hard" currencies are involved (U.S. dollar, British pound, German mark, French franc, Japanese yen), but even more so when "soft" currencies come into play. Exchange rates can fluctuate wildly, and governments can fix them at absurd levels that have little to do with their true value. And, of course, some are not convertible at all (the former Soviet Union's ruble, for example).

And then there are laws in some countries restricting the movement of money—sometimes you can make it there but cannot take it home. Even more challenging are "goods or services in kind" arrangements, where money is not used to pay at all—the barter system. These deals can get horribly convoluted.

Be certain to agree upon and specify the currency for payment, the exchange rate or method for determining it, the place of payment, timing, and any restrictions from any governments involved. Carry these understandings through your progress payment, retention, and change order clauses and incorporate them into your quotation, billing, and payment procedures.

Transportation and Shipping. Customs and immigration services come into play here, and their regulations can be unbelievably complex and arcane. These areas must always be investigated in advance. Some countries restrict certain materials, tax them heavily, require extensive documentation, or prohibit their importation entirely. There's little logic, in most cases, as to which they will do for what items.

It is not only foreign customs laws that one must know, but your own. If you are bringing in supplies or materials, your own country may restrict them or delay them somewhat. That is the operative word—delay. Many times the tax or fee (duty) is not so significant, but the documentation and procedural issues lead to impoundment, return, or destruction of what is shipped.

When I use the term "supplies" or "material," remember that these include design drawings, technical specifications, operating manuals, training materials, system software, and all the other "intellectual property" that makes up almost every project. These can be separately restricted by a whole host of other regulations and international agreements (copyright, patent, censorship). The duty on these and the delay they are subject to—the hassle, in other words—can be immense.

There is also the packaging and labeling of items. Something you would call "presentation boards" to train operating personnel might be interpreted by a South African custom official as "jewelry display cases"—a sensitive subject in a country wanting to protect its diamond industry. Little inconsistencies here can lead to huge costs and delays. Investigate the procedures and classifications, stick to them rigorously, and be prepared to work around any delays or difficulties—there is a good chance you'll have to.

Let's not forget the "shipment and transportation" of *people*. Some countries require work permits, visas, and/or residency permits. Some of these are good for only a certain period, and the holder must physically leave the country and reapply if he or she wants to continue working there. These always seem to expire right before the engineer or technician in question is scheduled for a major task. In Japan, for example, engineers with whom I was working were issued permits expiring in six months. Since the project lasted two years, they were sent back to the United States for a week or so, then readmitted under a renewed six-month permit. For project management this was a costly nuisance and a planning challenge. But for the American engineers involved, it was an unexpected benefit. They got a week's vacation in Hawaii every six months!

It is not uncommon for countries to require that certain tasks be performed by nationals—to prohibit using foreign architects, engineers, accountants, and the like, or to mandate that a certain percentage of the work be done by its citizens. Or a certain share of the contracting (or subcontracting, or materials, and so on) may be required.

Little items can stop big efforts. A project manager can be delayed in bringing her family to the job site because of laws dealing with pet quarantine. Passports expire at the most inopportune times, visas run out. The potential nuisances, or the potential for nuisances to become tragedies, is never ending. The point is to investigate these peculiarities, never take them for granted, and keep posted should they change in mid-project. And, of course, take the time involved into your consideration, and provide contingencies should delay occur. Always have a backup person ready to substitute should your key employee get snagged by one of them. And by all means make sure that they have valid papers.

Taxes. If you think taxes are complex and illogical in your country, try dealing somewhere else—they're obfuscating worldwide. Not only are there income taxes and sales taxes, taxes based on property valuations, and the like, there are taxes on people (head taxes), certain materials, foods, services, value-added taxes (VATs), and on and on.

There is also a patchwork of reciprocity issues. In Spain, for example, income taxes taken from a foreigner's wages can be 25 percent, but if that foreigner is an American (the United States has a reciprocity agreement with Spain), the rate is only 10%. But the lower rate applies only if certain registration and residency documents are submitted—in advance. Some countries issue tax credits or level surcharges for one reason or another (use of local labor, training of nationals, most-favored-nation trading agreements, and the like).

Variations in tax treatment can have a significant impact on wage rates, salaries, bonuses, and fringe benefits. Or on where you buy your materials and equipment—for that matter, on what type of material and equipment you specify or provide.

In this category, more than most, there is no substitute for up-to-date legal and tax counseling *before* you contract. And, as you'd expect, these rules and exceptions change constantly. Find out what they are, provide for them in your prices and in your written agreements and salary or wage negotiations, and keep up with their changes. And do not forget to account for local and regional taxes in your plans—and in your contracts.

Work Methods and Habits.   Americans are accustomed to a five-day work-week with weekends off. This does not apply in many countries. Nor do the work hours we prefer. In some places work begins at midmorning and continues until 10 or 11 in the evening, or later. We're also used to staggered vacations, where each individual chooses his or her time off. Not so in many countries, where, it seems, the entire nation takes vacation at the same time. And, of course, religious or political holidays vary widely. Schedules, prices, and practices need to be tailored to account for all these variations.

Then there are union rules and craft traditions, all subject to difference from location to location. Here an example is in order. As an owner's representative, I once contracted with an American firm to furnish and install a turbine generator at an electric station in the United States. The company bought the expensive equipment from the German manufacturer, and it was technically superb. The problem was that much of the wiring and piping that we are accustomed to installing at the job site was done in the German factory, eliminating a lot of work for our pipefitters and electricians. Traditional division of labor issues plagued the installation. When a complex subsystem of motor control centers, valves, piping, and electrical heaters came as a module, the electricians demanded to install it and so did the pipefitters. Maybe this didn't matter to the German manufacturer, but it sure mattered to the general contractor and the owner.

The examples are endless. Japanese contractors will have more inspectors and supervisors per construction worker than Americans will. Certain Middle Eastern workers will simply not do certain work—they consider it beneath them. The concept of time, urgency, what "soon" or "immediately" means is the subject of all sorts of misunderstanding and irritation as it varies from culture to culture.

Safety regulations and practices also vary, more than you would expect. Most oil firms operating in the North Sea were shocked to learn what the Norwegian government required for living quarters, recreational facilities, and safety precautions for offshore workers.

Do not presume as you go from the United States or a European nation to another country that safety or social conditions are lesser issues—they could be far greater. Local rules will cover such obvious areas as health, sanitation and fire protection, but also some not-so-expected, such as prayer facilities that must be provided for workers, health care benefits, maternity, and paternity leave, child care, food preparation, consumption or trade in prescription drugs, alcohol, and even certain foods. Periodicals, motion pictures, and videos are also subject to censorship or restriction.

Technical Requirements.   Conversions between the English and metric systems of measurement are pretty well understood today, and the trend toward universal metric systems is well under way. But this is an obvious complexity—one most are used to handling. There are not-so-obvious ones.

Testing standards vary from place to place, as do jurisdictions or traditions regarding the appropriate standards-setting institute. Americans are used to ANSI (American National Standards Institute) standards, but these do not apply everywhere. It's true that, in many fields, U.S. standards predominate, but this cannot be assumed. Neither can material compositions, such as the amount of carbon in high-strength steel or the chemical elements and water ratios of concrete. In addition, some countries regulate the amount of domestic material, or the percentage of domestic components, that must be used in certain cases.

Operating equipment, computer systems, engines, pumps, motors—anything that needs special tools to install or maintain, and most important, spare parts or consumable supplies—are a source of concern. Do not assume that you can go to the local hardware store or building supply house and get these. Check on availability and fitness before specifying, particularly with foreign-made or designed equipment.

Specifications formats differ, as do the roles and liabilities of certain design professionals regarding technical issues. Also keep in mind that technical specifications imply time and money, always. When these differ, even in seemingly insignificant ways, the cost and time for compliance can swing widely. Insisting on your specifications may be simple for you, as specifier, but terribly expensive to the cross-border party. Even fabrication and construction or installation decisions have subtle impacts. For example, if a component is assembled before importation, the duty can be 10 times what it would be if it was fabricated and imported in components that are assembled after it crosses the border.

Another general rule comes to mind here. Do not assume that the foreign national (the French architect, the Russian engineer, the Japanese scheduler) knows these differences! Each may know his or her "understoods," but they don't know they are different from yours because they are so "understood" to them.

The analogy of fish in water illustrates this subtlety. If you want to know about living in water, don't ask a fish—it doesn't know about living in air, so it cannot describe the difference. In fact, some suggest that fish don't even know what water is, or that it exists, because it surrounds them and is invisible and they have never lived outside it! That is a good description of many construction experts and contract managers—they have only lived in their framework of rules and procedures, so they cannot recognize how particular they are and how they might differ from others in other countries.

Contract Documents (and Underlying Law).   A barely adequate description of these differences would fill many volumes. Here I'll point out some major differences and give some general guidelines. Nothing here should take the place of competent legal counsel versed in the particulars.

Americans are familiar with the document structure and forms described in Chapter 5. Whether they are owner-provided, AIA forms, or some other standards, they may not be appropriate or acceptable for cross-border contracting. The most common documents for international contracting are those published by the FIDIC (Federation Internationale des Ingenieurs-Conseils, or in English translation, the International Federation of Consulting Engineers). This form is formally referred to as "Conditions of Contract for Works of Civil Engineering Construction" but applies to almost every type of construction work. They are commonly called the FIDIC forms, and a large body of international law has developed and continues to evolve around these terms and conditions. If you are doing work outside the United States, you should become familiar with these forms and their underlying case law. If you are dealing on a domestic project but using foreign material or nationals (or have foreign investors, contractors, and so on), you should expect that they would be familiar with the FIDIC. If you are using non-FIDIC documents, you should know the difference between them and FIDIC.

Contract terminology also differs, either in formal or conversational usage. For example, in the United States, AIA documents contain "changed conditions" clauses (AIA 1976 G.C. clause 12.2). In the United Kingdom, these are "unfavorable physical conditions" clauses. Americans use the term "change order," whereas the British use "variation orders." In the United States, proposals or bids are sent by contractor to owner to apply for contracted work. In Europe, the United Kingdom, and many other places, these are called "tenders."

Needless to say, what constitutes a dispute, delay, change (variation), acceleration, force majeure, and so on, varies from document to document and jurisdiction to jurisdiction. So do the process and the participants involved in dispute resolution, claims, final acceptance—even progress estimates and partial payments.

In the United Kingdom, for example, and in other countries where their influence has been felt (Australia, South Africa, Canada, and others), a large number of contracts are done on a "bill of quantity" basis, one very similar to the American unit price method, but not identical to it. *Quantity Surveyors,* licensed professionals, are employed to perform some of the functions described here for the contract manager. So not only is the work measured differently and paid for differently, the people involved in both are different. The roles and relationships, the liability for error, the final word in disputes—all these change to one extent or the other.

Commercial Terms, General Conditions.   The best way to assure that your documents and procedures reflect the circumstances of cross-border contracting is to undertake a thorough review of your (or their) documents from beginning to end. Pay particular attention to the feasibility and legality of terms dealing with warranties, bonds, insurance, liens, dispute resolution, labor relations, and bid advertisement or evaluation processes.

That's the best advice I can give in this regard: Never assume that your (or their) "boilerplate" is adequate or even enforceable. Take the commercial terms and conditions you plan to use and go right down the list of clauses, one by one, to verify that they make sense, are acceptable to the other party, and reflect the law and business customs under which you will be working.

Local Laws, Authorities, and Practices. Keep in mind that you may understand and have accounted for the laws of the sovereign nation of Japan but not the Shinagawa Prefecture. Or that you've got a handle on the Spanish construction market but not the state of Galicia. In other words, do not assume monolithic customs or uniform legal, commercial, or technical requirements within any given country. These vary in the United States from state to state and municipality to municipality. They vary just as much, sometimes even more, within other nations. Obvious areas of special attention here include taxes, safety and environmental regulations, working conditions, customs, fees, and labor relations.

Risk and Rewards. Cross-border contracting is, as you can expect, fraught with risk and uncertainty. The demands on contract management are intense, and the reliable "understoods"—the documents, players, laws, costs, and languages—are fewer and less dependable. But all these warnings aside, cross-border contracting is extremely profitable and satisfying when done correctly. Refreshing alternatives to buying, selling, or building present themselves. New perspectives are gained, new techniques and new methods are learned. And, as ever, the opportunity to gain deeper understandings of others, and to learn how much we are alike, is invaluable.

Regardless of how you view the risks and the potential rewards, cross-border contracting is on the increase and is certainly the wave of the future. So whether you are doing it now, as buyer or seller, you are bound to encounter it more and more as time passes. My advice is to welcome it—to take advantage of the opportunities it presents to you individually and to the success of your construction experience. But the most important benefit, the one I'd like to emphasize in summary, is that good cross-border contract management makes for good contract management—period. By forcing us to intensify our understanding of the laws, practices, rights, and duties embedded in contract management, it sharpens our understanding and increases our efficiency and control. It makes us consider risk, and provide for it. It helps us uncover and discard outdated assumptions and obsolete principles. In short, it helps us improve. It makes us more alert and prepared, more competent in our contract management, and more confident as we move ahead in our project, in our profession, and in our careers.

GLOSSARY OF TERMS

ADDENDUM: Document issued by the owner to bidders prior to the execution of the contract that modify or interpret the bidding documents (request for proposal), including drawings, by additions, deletions, clarifications, or corrections. These changes should become incorporated into the contract documents prior to award of the contract.

ADDITION TO CONTRACT PRICE: Amount added to the contract price by change order. *See also* Extra.

ADMINISTRATION, CONTRACT: The management of commercial transactions required for contractual compliance. Begins with contract award and ends with contract closeout or termination.

ADVERTISEMENT FOR BIDS: Published public notice soliciting bids for a construction project. Most frequently used to conform to legal requirements pertaining to projects constructed under public authority, and usually published in newspapers of general circulation in those districts from which the public funds are derived.

AGENT: One authorized by another to act in his or her stead or behalf.

AGREEMENT: (1) A meeting of minds. (2) A legally enforceable promise or promises between two or among several persons. (3) The specific contract document stating the essential terms of the construction contract and incorporating by reference the other contract documents.

AGREEMENT FORM: A document setting forth in printed form the general provisions of a contract, with spaces provided for insertion of specific data relating to a particular project. It is a part of the request for proposal sent to bidders.

ALTERNATE BID: Change in project scope, proposed pricing method, or alternate materials and/or methods of construction offered by a bidder in lieu of, or in addition to, the solicited base bid.

APPLICATION FOR PAYMENT: Contractor's written request for payment of amount due for completed portions of the work.

APPROVAL, ARCHITECTS OR ENGINEER'S: Architect's (engineer's) written or imprinted acknowledgment that materials, equipment, or methods of construction are acceptable for use in the work, or that a contractor's request or claim is valid.

ARBITRATION: Method of settling claims or disputes between parties to a contract, rather than through litigation. An arbitrator or a panel of arbitrators, selected for their specialized knowledge in the field in question, hear the evidence and render a decision. Parties must agree to arbitrate, and arbitrators' rulings are usually enforceable by law.

ARCHITECT-ENGINEER (A–E): An individual or firm offering professional services as both architect and engineer; used to refer to the company responsible for designing a facility.

AS-BUILT DRAWINGS: Drawings of the contracted project, or portion thereof, showing the actual condition of the project elements at completion. They can be marked construction drawings, showing deviations from planned work, or separately drafted drawings. Submission of required as-built drawings is usually a prerequisite to a contractor's final payment.

AUDIT: An examination of a contract activity, controls, or charges to determine if they are correct, effective, prudent, and/or adequate.

AWARD: A communication from an owner accepting a bid or negotiated proposal. An award is the final step in the contract formation process.

BACKCHARGE: Process or document by which the owner charges a contractor or vendor for the cost of correcting work that the contractor or vendor failed to perform or performed incorrectly.

BASE BID: Scope of work and corresponding amount of money stated in the bid as the price for which the bidder offers to perform the work. Usually refers to a bid that conforms to the RFP requirements.

BID: A complete and properly signed proposal to perform the work or designated portion thereof for the prices stipulated therein. Supported by submittals required for the RFP. Also called proposal.

BID BOND: A form of bid security executed by the bidder as principal and by a surety. The surety protects the owner from damages should the bidder fail to sign a contract to perform the work described in the RFP if the owner accepts its bid.

BID DATE: The date established by the owner for the receipt of bids.

BID SECURITY: The deposit of cash, certified check, cashier's check, bank draft, money order, or bid bond submitted with a bid and serving to guarantee to the owner that the bidder, if awarded the contract, will execute such contract in accordance with the terms of the RFP.

BILL OF QUANTITY: (1) Type of contract common in the United Kingdom and other countries. Similar in payment method and process to the unit cost method. (2) Table listing payment items and prices.

BOILERPLATE: Term informally used to represent the collection of standard legal and commercial terms, conditions, and clauses used for all contracts for a project, such as general and special conditions.

BONDING CAPACITY: The total dollar value of contracts that the surety will guarantee for a contractor.

BONUS AND PENALTY CLAUSE(S): Provisions in the contract for payment of a bonus to the contractor for completing the work prior to a stipulated date, and/or for a charge against the contractor for failure to complete the work by such a date. These clauses are also used as incentives for specified cost and technical performance.

CERTIFICATE FOR PAYMENT: A statement from the engineer, construction manager, or other owner's representative to the owner confirming the amount of money due the contractor for work accomplished.

CERTIFICATE OF INSURANCE: A memorandum issued by an authorized representative of an insurance company stating the types, amounts, and effective dates of insurance in force for a contractor. The industry-wide "accord" form is generally used.

CHANGE ORDER: A written order to the contractor signed by the owner or its agent (such as the A–E, or construction manager), issued after the execution of the contract, authorizing a change in the work or an adjustment in the contract price or the contract time. A change order may be signed by the owner's agent alone, provided the agent has authority from the owner for such procedure and that a copy of such written authority is furnished to the contractor upon request. A change order should also be signed by the contractor signifying its agreement to the adjustment in the contract price or the contract time. The contract price and the contract time should be changed only by change order.

CHANGES IN THE WORK: Changes ordered by the owner consisting of additions, deletions, or other revisions within the general scope of the contract, the contract price, and the contract time being adjusted accordingly. All changes in the work should be authorized and documented by a change order.

CLAIM: Request from a contractor for adjustment to the contract price, contract time, and/or contract requirements.

CLOSED SPECIFICATIONS: Specifications stipulating the use of specific products or processes without provision for substitution.

CLOSEOUT: The process of formally terminating a construction contract upon completion of the work or termination of the agreement because of other reasons (such as contract breach or default or cancellation of the contract).

CODES: (1) Regulations, ordinances, or statutory requirements of a governmental unit relating to building construction and occupancy, adopted and administered for the protection of the public health, safety, and welfare. (2) Standards of design, construction, materials, tests, or other products or methods established by industry or governmental bodies and often cited or referenced in the specifications.

COMPLETION DATE: The date established in the contract documents for substantial completion of the work.

CONDITIONS OF THE BID: Conditions set forth in the instructions to bidders, the notice to bidders or advertisement for bids, the invitation to bidders, or other similar documents prescribing the conditions under which bids are to be prepared, executed, submitted, received, and accepted.

CONDITIONS OF THE CONTRACT: Those portions of the contract documents that define, set forth, or relate to contract terminology; the rights and responsibilities of the contracting parties and of others involved in the work; requirements for safety and for compliance with laws and regulations; general procedures for the orderly prosecution and management of the work; payments to the contractor; and similar provisions of a general, nontechnical nature. The conditions of the contract typically include general conditions and special or supplementary conditions.

CONSTRUCTION MANAGER: (1) Special management services performed by a contractor, the A–E, or others during the construction phase of the project, under separate or special agreement with the owner, or by the owner itself. Generally, this is not part of the contractor's or engineer's basic services but is an additional service sometimes included. (2) Designated individual serving as highest owner authority during the project construction period.

CONSTRUCTIVE CHANGES: *See* Informal changes.

CONTRACT: A legally enforceable promise or agreement between two or more parties.

CONTRACT DOCUMENTS: The agreement, the conditions of the contract (general, special, and other conditions), the drawings, the specifications, all addenda issued prior to execution of the contract, all modifications thereto, and any other item specifically stipulated as being included in the contract documents.

CONTRACT PRICE: The price stated in the owner–contractor agreement, which is the total amount payable by the owner to the contractor for the performance of the work under the contract documents.

CONTRACT TIME: The period of time established in the contract documents within which the work must be completed.

CONTRACTOR: (1) One who contracts. (2) In construction terminology, the person or organization responsible for performing the construction work and identified as such in the construction contract.

COST-PLUS-FEE AGREEMENT: An agreement under which the contractor is reimbursed for its direct and indirect costs and, in addition, is paid a fee for its services. The fee is usually stated as a stipulated sum or as a percentage of cost.

CRITICAL PATH METHOD (CPM): A planning and scheduling technique involving the charting of all events and operations to be encountered in completing a given process, rendered in a form permitting determination of the relative significance of each event and establishing the optimum sequence and duration of operations.

DATE OF COMMENCEMENT OF THE WORK: The date established in a notice to proceed or, in the absence of such notice, the date of the agreement or such other date as may be established therein or by the contracting parties.

DATE OF SUBSTANTIAL COMPLETION: The date when the work or a designated portion thereof is sufficiently complete, in accordance with the contract documents, so the owner may occupy the work or designated portion thereof for the use for which it is intended.

DEFAULT: Contractor or owner's failure to perform under the contract.

DEPOSIT FOR BIDDING DOCUMENTS: Monetary deposit required to obtain a set of request for proposal documents. Customarily, this is refunded to the bona fide bidders on return of the documents in good condition within a specified time.

DESIGN–BUILD: Use of a single firm to both design and construct a facility.

DRAWINGS: The portion of the contract documents showing in graphic or pictorial form the design, location, and dimensions of the elements of the work to be constructed. Sometimes called "plans."

ENGINEER: Designation reserved, usually by law, for a person or organization professionally qualified and duly licensed to perform engineering services.

ESCALATION: A change in cost over time, either a decrease or increase, resulting from a change in the price of a particular element from the bid and/or contract price.

ESTIMATE, CONTRACTOR'S: A forecast, as opposed to a firm proposal, of construction cost, prepared by a contractor for a project or a portion thereof. (2) A term sometimes used to denote a contractor's application or request for a progress payment.

EXTRA: An item of work not included in the contractual scope, which may involve additional cost and time.

FAST TRACKING: (1) Method of design and construction whereby a series of tasks, jobs, or contracts are overlapped to reduce the length of time required to complete the project. (2) The overlapping of design and construction phases so that construction begins before design is completed.

FINAL ACCEPTANCE: The owner's acceptance of the work from the contractor when satisfied that it is complete and in accordance with the contract requirements. Final acceptance precedes final payment unless otherwise stipulated.

FINAL INSPECTION: Final review of the contractor's work by the architect, engineer, construction manager, or owner prior to final acceptance and final payment.

FINAL PAYMENT: Payment made by the owner to the contractor of the entire unpaid balance of the contract price as adjusted by change orders, back-charges, or other amendments.

FORCE ACCOUNT: Construction by owner.

FORMAL CHANGES: Written directives from the owner to a contractor to change the time, cost, or scope of work from that originally specified by the contract documents.

FORMATION, CONTRACT: The series of tasks and activities leading to achievement of a written contract signed by both principal contracting parties.

FRONT-END LOADING: The practice by bidders of increasing the bid amount of items or work to be performed early and reducing the price of items to be performed later so as to maximize initial progress payments.

GENERAL CONDITIONS OF A CONSTRUCTION CONTRACT: General requirements that apply to all construction contracts for a particular project. The basic relationships between the contracting parties, general project rules, and commercial terms are contained in this section.

GENERAL CONTRACTOR: The function of the general contractor is to coordinate the activities of the various parties and agencies involved with the construction and to assume full, centralized responsibility for the delivery of the finished product within the specified time.

GUARANTEED MAXIMUM COST: Amount established in some cost-plus contracts as the maximum cost of performing specified work on the basis of cost of labor and materials plus overhead expense and profit. Unless modified by changes, the owner's payment obligation does not exceed this amount.

INFORMAL CHANGES: Changes to a contractor's scope of work, cost, or required time of performance brought about by acts of God, acts or omissions of the owner or others, or circumstances beyond the control of the contractor.

INVITATION TO BID: A portion of the bidding documents soliciting bids for a construction project for which bidding is not open to the public.

JOINT VENTURE: A collaborative undertaking by two or more persons or organizations for a specific project or projects, having the legal characteristics of a partnership.

LABOR AND MATERIAL PAYMENT BOND: A bond of the contractor in which a surety guarantees to the owner that the contractor will pay for labor and materials used in the formation of the contract. The claimants under the bond are defined as those having direct contracts with the contractor or any subcontractor.

LETTER OF INTENT: A letter or other communication issued upon award to the successful bidder, signifying an intention to enter into a formal agreement, usually setting forth the general terms of such agreement by referencing the request for proposal. Letters of intent are sometimes issued between contract award and signing of the conformed contract.

LIQUIDATED DAMAGES: Sum of money, usually given as dollars per day, stipulated in a construction contract as a penalty against the contractor for failure to complete the work by a designated time.

LUMP-SUM CONTRACT: Contract in which a specific amount is set forth as the total payment for performance of the contract.

MECHANIC'S LIEN: A lien on real property created by statute in all states in favor of persons applying labor or materials for a building or structure for the value of labor or materials supplied by them. In some jurisdictions, a mechanic's lien also exists for the value of professional services. Clear title to the property cannot be obtained until the claim for the labor, materials, or professional services is settled.

MILESTONE: (1) A key project event whose accomplishment is scheduled and against which progress is closely measured. (2) A specific achievement designated in a contract progress payment arrangement.

MULTIPLE PRIMES: The use of several prime contractors, each under separate contract with the owner, to perform specific construction work.

NOTICE TO PROCEED: Written communication issued by the owner to the contractor authorizing it to proceed with the work and establishing the date of commencement of the work.

NOTIFICATION: Transmitted from the owner to a contractor informing it of anticipated changes in the work and requesting a response indicating the cost and schedule impact of the changes. Often serves as a transmittal for revised drawings or specifications.

PAROLE EVIDENCE RULE: A rule of law that disallows oral evidence from being introduced to contradict what the parties have agreed to in writing.

PERFORMANCE BOND: A bond of the contractor in which a surety guarantees to the owner that the work will be performed in accordance with the contract documents. Except where prohibited by statute, the performance bond is frequently combined with a labor and material payment bond.

PREQUALIFICATION OF PROSPECTIVE BIDDERS: The process of investigating the qualifications of prospective bidders and selecting those who will be invited to bid, on the basis of their technical competence, financial integrity, and past performance.

PRIME CONTRACTOR: Any contractor on a project having a contract directly with the owner (as opposed to a subcontractor).

PROGRESS PAYMENT: Partial payment made during progress of the work on account of portions of the work completed. Usually made on a monthly basis.

PROJECT MANAGER: The individual representing the owner who has overall responsibilities for a project.

PUNCH LIST: A list of uncompleted or corrective items of work to be done to complete the contract work.

QUALIFIED BIDDER: Contractor who has been allowed by the owner to bid when bidding is by invitation.

REIMBURSABLE EXPENSES: Amounts expended for or on account of the project that, in accordance with the terms of the appropriate agreement, are to be reimbursed by the owner.

RELEASE OF LIEN: Instrument executed by one supplying labor, materials, or professional services on a project that releases the individual's lien against the project property. Usually, a prerequisite to final payment.

REQUEST FOR PROPOSAL (RFP): The advertisement or invitation to bid, instructions to bidders, the bid (proposal) form, and the proposed contract documents, including any addenda issued prior to receipt of bids.

RESPONSIVE BID: Bid submitted in good faith, properly signed, complete, and in prescribed form that meets the conditions of the bidding requirements.

RETAINAGE: Also called retention. A sum withheld from progress payments to the contractor in accordance with the terms of the construction contract. Traditionally, it is 10% of payments due the contractor and payable upon final acceptance of the work.

RIPPLE EFFECT: A change in one area, or to one contractor, that causes changes in other areas of the project or to other contractors.

RISK PROFILE: A ranking or estimate of risk inherent to a contract type or contractor. Used to target contractual controls or audit approaches.

SHOP DRAWINGS: Drawings, diagrams, illustrations, schedules, performance charts, brochures, and other data prepared by the contractor or any subcontractor, manufacturer, supplier, or distributor that illustrate how specific portions of the work shall be fabricated and/or installed. They are submitted for the owner's approval prior to starting the affected work.

SHORT-FORM (FIELD) CONTRACT: Construction contract prepared and awarded by field office for short-term, low-cost work, not in the scope of present or planned construction contracts.

SPECIFICATIONS: The technical portion of the contract documents consisting of written descriptions of materials, equipment, construction methods, standards, and workmanship.

SPREADSHEET: Form used to record results of a bid evaluation.

SUBCONTRACT: Agreement between a prime contractor and a subcontractor for a portion of the work for which the prime contractor is under contract with the owner to provide.

SUBCONTRACTOR: A person or organization who has a direct contract with a prime contractor to perform a portion of the work.

SUB-SUBCONTRACTOR: A person or organization who has a contract with a subcontractor to perform a portion of the work.

SURETY BOND: A legal instrument under which one party (surety) agrees to answer to another party (owner) for the debt, default, or failure to perform of a third party (contractor).

TAKEOFF: Process whereby a cost estimator determines required resource quantities and applies applicable rates to obtain a cost estimate.

TIME AND MATERIALS (T AND M) CONTRACT: Contract that provides the contractor with reimbursement for all direct labor hours at a fixed hourly rate and for the actual costs of material used. The contractor's overhead and profit are embedded in the hourly rate.

TENDER (DOCUMENTS): *See* Bid.

TURNKEY CONTRACTOR: A contractor responsible for all phases necessary to complete design and construction of a facility. Theoretically, once the owner has expressed its needs, the turnkey contractor does everything up to and including turning over the facility's "key" to the owner.

UNBALANCED BID: A deliberate practice by bidders of weighing or factoring component bid amounts to: (1) front-end load; or (2) increase unit prices for additions that the bidder expects to occur in significant quantities or decrease unit prices for items that the bidder feels will not be ordered in large quantities or may later be deleted from the scope of work.

VARIATION ORDER: *See* Change Order.

VENDOR: Supplier of material, equipment, or finished goods to be used in construction without labor and supervision for their installation or erection.

WAIVER OF LIEN: An instrument by which a person or organization who has or may have a right of lien against the property of another relinquishes such right. In a few jurisdictions, this right can be waived even before the construction required by the contract documents.

WORKMEN'S COMPENSATION INSURANCE: Insurance covering liability of an employer to his employees for compensation and other benefits required by workmen's compensation laws with respect to injury, sickness, disease, or death arising from their employment. Required by law in each state of the United States.

INDEX

Addenda:
  definition of, 77
  sample, 122–123
  use, 121
Addition, *see* Change orders
Administration, contract, 9, 18
Advertisement, 85
Agreement:
  contract document, 72, 94
  sample, 107
Arbitration, 211
Auditing, contract, 11, 247
Audits, contract:
  clauses, 256–257
  cost, 248
  financial, 248
  operational, 251
  techniques for, 248–254

Backcharges, 10, 197
  cost collection, 197
  procedure for, 199
  sample form, 200
  summary report of, 244
Bid:
  cycle, 8, 119
  error, 119, 135
  receipt and evaluation, 9, 129
  security, 131
  shopping, 141
  unbalancing, 93

Bidding:
  complementary, 119
  error, 119
  extensions, 124
  uncertainty, 120
Bid packages, contents, 72
Bonds, 78–79
Bonus/penalty clauses, 40, 55
Bulletins, *see* Notifications

Ceiling Price, 46
Changed Work:
  documents, 186
  exceptions, 193
  history form, 192
  process, 183
Change orders, 77, 179
  approval form, 187
  log, 201
  sample form, 188
  short form versions, 220
  summary report of, 243
Changes:
  causes, 180
  constructive, 179
  formal, 179
  informal, 179
  types, 179
Claims, 11, 203
  analysis of, 208
  causes of, 204

delay, 209
elements of, 205
evolution of, 213
overhead, 210
procedure for, 212
structure of, 207
summary report sample, 242
Closeout, contract, 11, 223
documentation for, 223
sample checklist, 227
suggested procedure for, 226
typical deliveries involved, 223
Commercial pressures, 2
Contract:
administration, 9, 18, 145
ancillary types, 78
auditing, 247
award, 9, 138
claims, 202
closeout, 223
conformance, 142
documents, 8, 17, 69, 94
equipment rental, 78
evaluation form, 229
formation, 7, 67
identification, 65
monitoring, 11, 233
packaging and scheduling, 7, 58
planning, 5, 15
pricing, 7, 36
records, 152
reporting, 11, 235
short forms, 78, 216
strategies, 25
Contracting elements, 4
across borders, 275
Contract strategies, 25
Controls:
operational, 3
Cost, contract, 37
Cost plus pricing, 51
Cost reimbursable, 51
Costs:
contractor's, 37
evaluated amounts, 131, 132
life cycle, 131, 132
nontechnical, 131, 132
payments based on, 158
Critical path scheduling, 60

Design build, 26
Documents, contract, 8, 69
for ancillary contracts, 78
development of, 69

hierarchy, 75
industry standards, 70
language considerations for, 80
modifications to, 77
owner reference sets, 69
specimen versions, 98
Drawings:
as-builts, 224
contract, 73, 95

Escalation, 40, 45
Extra, *see* Changes

Fair price estimate:
definition, 133
request form sample, 186
Fast tracking, 32
Fee, contractors, 38
Few primes, 30
FIDIC contracts, 281
Firm fixed price, 43
Fixed price incentive, 46
Force account, 33
Formation, contract, 7, 67

General conditions, 72, 91, 94, 112
General contractor, 29
Guaranteed maximum pricing, 46

Incentives:
cost, 54
performance, 40, 54–55
profit, 54
schedule, 55
technical, 54
Incremental cost method (for pricing claims), 208
Insurance:
certificates, 150
coverage, 79
log, 150
Invitation (as bidding process), 85
Invitation to bid, 72, 92, 98

Labor broker, 33
Lawyer's role, 261
Letter of intent, 142
Liquidated damages, 55
Lump sum:
with escalation, 45
pricing, 43

Mobilization:
contract commencement, 9, 147

of contractor, 60
meetings, 151
Multiple primes, 32

Negotiation, 139
Notifications:
log, 191
of proposed changes, 183
sample form, 184
of termination, 227

Operating and maintenance manuals, 225
Organizational strategies, 6, 23

Parole Evidence rule, 271
Performance monitoring, 19
Planning, contract, 15
Post performance evaluation, 226
Prebid conferences, 125
Preliminary scope determination, 59
Price, contract, 38
Pricing:
alternatives, 7, 39, 43
of change orders, 56
contractor perspective, 42
proposal structure for, 92
subtier, 41
terms, 37
Progress billings and payments, 10, 157
alternative bases, 157
estimate forms, 170–171
final payments, 172, 226
objectivity in determining, 160
recommended procedure for, 167
Progress measurement:
reasons for, 160
techniques used, 162–166
Project organizations, 6, 23
Project orientation, 24
Proposal:
form, 72, 92
sample, 100
Propriety, 125
Purchase orders, 76, 77

Qualification:
of bidders, 8, 85
suggested procedure for, 87
timing of, 88
Quantity Surveyor, 281
Questionnaires, bidder, 87

Release of liens, 223
Reporting:
contract, 235
hierarchy, 238
key indicators used, 245
sample reports, 240–244
Requests for proposals:
contents, 72
issuance of, 8, 91
Retainage, 172
Retention, 172
alternatives to, 173–174
Rights & Responsibilities, 2
Risk Profile, 254–255

Scheduling (of contract formation), 60
Scope of work, 59, 62–63
Short form contracting, 10, 217
change orders, 220
procedure for, 219
Site visits, 125
Special conditions, 72, 94, 114–118
Submittals:
bidder, 93
checklist, 149
contractor's, 148

Target:
cost, 46, 51
fee, 46, 51
formula, 46
labor hours, 51
price, 46
Technical specifications, 73, 94
Time and materials pricing, 53
Total cost method (for pricing claims), 208
Turnkey contract, 26

Unit Pricing, 44